Inside Japanese Ceramics

Inside Japanese Ceramics

A Primer of Materials, Techniques, and Traditions

Richard L. Wilson

Weatherhill
New York & Tokyo

WEATHERHILL
An imprint of Shambhala Publications, Inc.
2129 13th Street
Boulder, Colorado 80302
www.shambhala.com

First edition, 1995
First paperback edition, 1999
Sixth printing, 2022

Printed in China
♾ This edition is printed on acid-free paper that meets the American National Standards Institute Z39.48 Standard.
♻ Shambhala makes every effort to print on recycled paper.
For more information, please visit www.shambhala.com.
Weatherhill is distributed worldwide by Penguin Random House, Inc., and its subsidiaries.

Library of Congress Cataloging in Publication Data:
Richard, Wilson L. 1949–
Inside Japanese ceramics: a primer of materials, techniques, and traditions / Richard. L. Wilson
p. cm.
Includes bibliographical references and index.
ISBN 978-0-8348-0346-6 (hardcover)
ISBN 978-0-8348-0442-5 (paperback)
1. Pottery, Japanese—Handbooks, manuals, ect. I. Title.
NK4167.W56 1995
738'.0952—dc20 95-22235
CIP

Contents

Dedicated to potters and their teachers

Foreword

History, geography, and cultural practice divide the ceramics of Japan from those of the West. Americans and Europeans view our pottery as exotic, and one could hardly expect a different attitude. What cause me concern, however, are the unnecessary distortions, such as the romantic or idealist accounts of Raku ware or folkware, that displace other important parts of the tradition. The understanding of Japanese ceramics techniques outside Japan also suffers from selective biases: many Western potters seem to think that making Japanese ceramics is nothing more than haphazardly combining groggy clays and runny glazes with wood firing. These stereotypes are in need of correction.

Richard Wilson takes up this task in this book, and he is uniquely qualified for the job. Not only has he researched ceramic techniques throughout Japan, he is a potter himself. When he was a graduate student at our Kyoto City University of Fine Arts, he immersed himself in Japanese traditional techniques and materials. He pored over handwritten potter's manuals from the eighteenth and nineteenth centuries and traveled far and wide to dig up materials that even our own potters had forgotten about. He has fired unglazed stoneware with rural potters and painted overglaze enamels in ateliers here in Kyoto. Due to his breadth of knowledge and experience, the entirety of the Japanese technical tradition is presented here with an objective, even-handed viewpoint. The book begins by exposing the full panorama of Japanese ceramic history and proceeds with a chapter-by-chapter demonstration of the various techniques that grew out of that history. The reader is provided with a basic, systematic, and accurate approach to Japanese traditional methods, from clay preparation to firing.

As evidenced in all of the techniques in this book, traditional Japanese ceramics reflect a dedication to bringing nature to life. This attitude places severe restrictions on personal artistic freedom, but according to our practice this very restriction is the necessary precondition for bringing out the aesthetic qualities particular to the ceramic medium. Accordingly, the potter accepts the centrality of nature; self-expression and personal effort do not die, but rather are dedicated to finding a way to allow shape, color, and line to follow nature. This spirit of letting the material live has become the root of Japanese ceramics. The idea of giving material life according to its own wishes rather than willfully forcing it into something artificial is also connected with an ancient Japanese reverence for nature and with a philosophical way in which the attainment of selflessness is a goal. Craftspeople of the past used this spirit as a guide to making better products. Their sense of beauty was embodied in fine vessels, many of which survive to this day.

How does one master these techniques? Training customarily begins with following the methods of the great potters and recreating their classic works. But this is only a means for accomplishing the more important tasks of learning the properties of water, clay, and fire through physical experience and observing the resulting interaction of nature and person. This fosters a keen sensitivity to the essentials of the potter's art, one that becomes sharper with each encounter. The path leads to a deep, solid foundation from which original and creative work springs. This process, more than any canonical teaching, is the way tradition renews itself.

It is naive to think that the Japanese pottery tradition can be absorbed in America or Europe without a certain amount of misunderstanding and resistance. But we live in an era in which our destinies are increasingly intertwined and where it is ever more important to recognize the positive contributions of every civilization. Ceramics already reflect this truth and I believe that in the years to come they will do so to an even greater extent. In that regard, the significance of this book should be clear. People everywhere are beginning to see that the Japanese heritage is not unique or autonomous, but rather part of a greater legacy—that of humankind. Richard Wilson's book serves to underscore that fact. I hope that it encourages all potters to expand both their inner and outer horizons.

Kuriki Tatsusuke
Professor, Ceramics Department
Kyoto City University of Fine Arts

Preface

In terms of its technical and stylistic diversity, its expressive content, and the level of appreciation that it enjoys, the Japanese ceramics tradition is without compare. It has captivated me since I first picked up Herbert Sanders' *The World of Japanese Ceramics* in 1973. Since then, I have been gathering information on Japanese pottery techniques at every chance. The sources come from all over: interviews with many Japanese potters, concentrated study with a few of them, and over two years of studio work at the Kyoto City University of Fine Arts, an institution that has trained some of Japan's most renowned ceramic artists. My information has come to include some old potter's manuals too, even some found buried in Western libraries. These studies convinced me that our English-language material was still meager. Furthermore, something was needed to span the gap between historical and ethnographic accounts of Japanese ceramics, such as descriptions of life in a pottery village, and actual workshop practice.

The Japanese craftsperson accepts tradition as source of enormous power, both in an aesthetic and a technical sense. Classical standards of beauty and use are seen not as fetters but as guideposts on a road to self-development. Traditional materials and methods continue to have great importance, underlying not only the production of functional ceramics in small workshops but the making of much avant-garde work as well. To locate this tradition, however, one has to go beyond antiquarian delvings and trivia like secret glaze formulas or firing schedules. At the heart of tradition are basic procedures: methods of clay preparation, forming, decoration, glazing, and firing. A careful look at those procedures reveals a special efficiency, in which expressive effects develop out of the materials and methods themselves. This sensibility, I believe, is important for makers, users, and admirers of pots everywhere.

This book concentrates on tools, materials, and procedures and how they have informed the way traditional pots look and feel. The focus is on the major methods, ones that would be taught in most Japanese workshops. I have avoided the two extremes of arcane local variations and their terminologies on one hand and procedures already familiar to most Western potters on the other, and have attempted to find a balance between time-tested techniques and contemporary feasibility. The methods of traditional Japanese ceramics can provide practical studio solutions, and, beyond that, a model of integration and wholeness that may help the Western potter cope with the hard edges of life in the crafts. Above all the book is an invitation to participate, to study, make, touch, and use pots. I hope that readers will move from there toward a personal synthesis, taking into account national or regional lifestyles, local materials and economies. The bridge between the pottery village and the global village is ours to build.

Since most Japanese potters are reluctant to talk about technique, researching and writing this book took a long time. But I want to acknowledge those who did share information, both in word and deed. I am grateful to two master potters: Raku-ware maker Ogawa Chōraku, who has welcomed me to his Kyoto workshop since my very first visit to Japan; and Takatori-ware potter Kamei Rakuzan, who accepted me as his student back in 1977. Rakuzan's manner of teaching has been like Takatori ware itself, terse and compact. But what a story he tells with his hands, blunted from decades of exacting work! Rakuzan's pots have supported my belief in the possibility of beautiful, functional, and affordable stoneware.

At the Kyoto City University of Fine Arts (Kyoto Geidai), where I studied for two-and-one-half years, I received much help from potter-teachers Koyama Kihei and Kuriki Tatsusuke. It is a great honor that Kuriki, one of the finest potters in his generation, has written a foreword to this book. A teacher of Japanese technical history and glaze technology, Ōnishi Masatarō, was also a great inspiration. A more complete repository of technical history and lore is nowhere to be found. Thanks is also owed to Kyoto Geidai graduate Eguchi Akira, now chief teacher at Kyoto Craft Technical High School; his pottery primer was a very important source of ideas.

The production of the book proper was greatly assisted by three of my students at Rice University: Don Choi, who provided most of the drawings; Yun-Jung Lee, who carefully printed many of the photographs; and Bill Thompson, who proofread several drafts.

Photographer-colleague Geoff Winningham provided technical hints on the studio photography. Kim Schuefftan contributed countless ideas about the goals, contents, and presentation of this book, and potter-teacher John Neely, himself an avid student of Japanese techniques, provided a critical reading of the text. My original plan was to include a chapter on the trials and triumphs of non-Japanese potters who have worked in Japan, only to find it deserved separate treatment. I am nevertheless grateful for the insights provided by many Japan veterans, including potters worldwide—Janet Leach in England, Sophie Melon in France, Liou Chen-chou in Taiwan, and Michael Marcus, Miriam Friedman, Richard Bennet, and Rob Barnard in the United States. Publishing professionals in Tokyo who encouraged the project include the late Kanai Katsutoshi and his protégé Kuroiwa Teiji.

A principal collaborator in this book is Ogasawara Saeko. In addition to her own considerable expertise in calligraphy and the tea ceremony, she studied ceramics in Kyoto with Ogata Shūhei. She modeled for the photographs of the teaching procedures, helped with translating documents, and made important suggestions on the design. Most importantly, she has been a tower of encouragement, and I could never have completed the book without her help. Sketches of traditional potters at work are adapted from *Tōjiki Setsu* (Pottery and Porcelain Explained), originally published by Kyoto Prefecture in 1872.

Use of Japanese terminology in the text has been confined to words that have no simple equivalents; all other Japanese terms have been placed in an English-Japanese index. Macrons have been omitted for Japanese words in common English use like daimyo and shogun. Japanese personal names are given in traditional order, with family name first; premodern artists are referred to by their pseudonyms alone after first mention (e.g., Nonomura Ninsei is shortened to Ninsei).

ONE

Of Craft and Life

It is early morning, and the clay has been kneaded into neat balls, ready for throwing on the potter's wheel. A ball is slapped into place on the wheelhead, and the flywheel set into motion. Repeated raising and lowering of the clay mass achieve a solid center and provide a rhythm for the day's work. At the same time, the mind's eye opens to a favorite place: a valley hemmed in by forested hills (fig. 1). On the fringes of a village, where the terrain becomes too steep for house or paddy, kiln roofs jut into the sky, terminating in sooty chimneys. Pots are everywhere—new ones in the showrooms on the main street and old cracked ones, filled with flowers and miniature trees, in the back alleys. At the bottom of a shallow river, glazed potshards glimmer. This is a Japanese pottery village. Such places are separated by distance, historical circumstance, and not least of all, the ground underfoot. Traveling to a pottery village with one of its veteran makers, one hears how the shape of this mountain or the water level in that paddy hint at clay underneath. One is told how land and sea routes brought in outside influence, or how mountains prevented it. But most of all, it is the pots themselves that speak.

At first glance the products of each village seem unrelated, and at first encounter, the variety of shapes, decorations, and finishes from these many kilns is vast and confusing. With time, however, the pots come to tell a common and rather direct story, the story of their manufacture. Objects from the earth, they are earthlike. Pigments flash metallic impurity as much as they do color. Marks of the fire are counted as blessings, not blemishes. Japanese ceramics state the circumstances of their making, and that is what this book is about.

Japan is now a highly developed industrial country whose artistic horizons are as diverse, unfettered, and global as those of the modern West. But still there is space for traditions of all kinds, including ceramics, not only to survive but to prosper. The history and beauty of these traditional ceramics are inseparable from the way their makers work. This primer of traditional techniques is

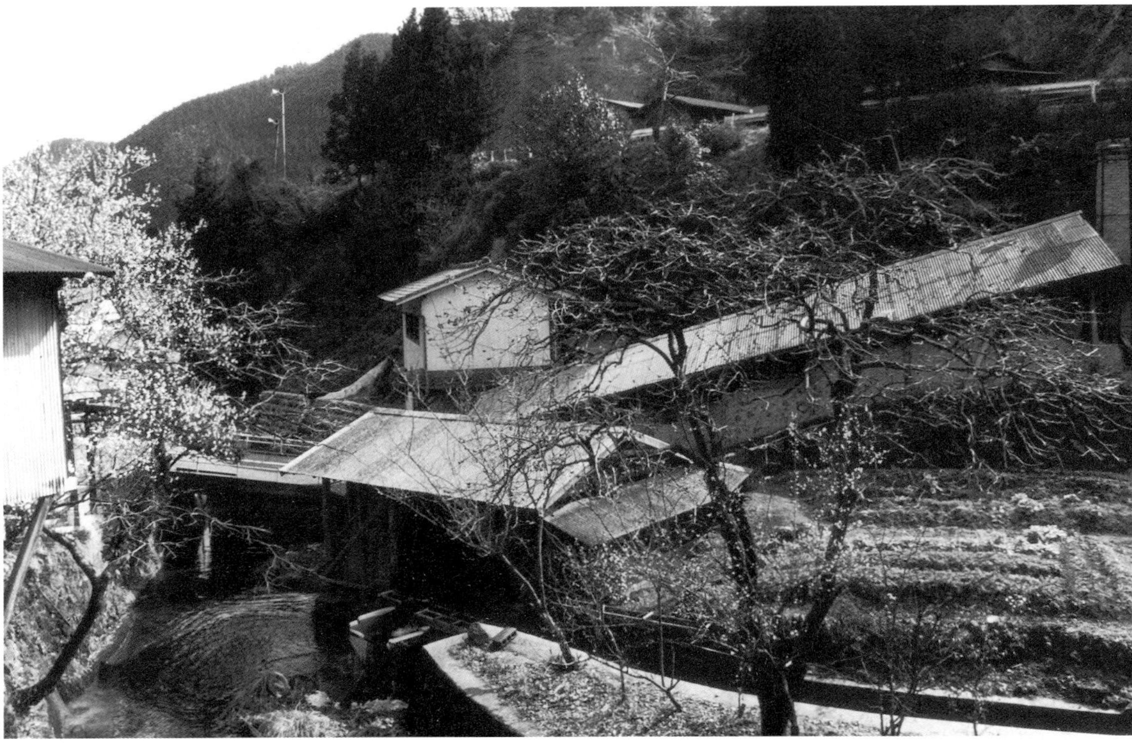

1. A Japanese pottery village

intended to demonstrate the achievements of the Japanese way and to generate both critical and creative views about today's Western approaches to the potter's art. This survey of Japanese techniques is also intended to serve as a means for the reader to achieve local and personal solutions. Pottery is nothing if not alive for both maker and user, and thus one may turn to Japanese techniques as to a foreign language—perhaps at first as an exotic fascination, but eventually as a means of communication, both informative and expressive. Therein lies the real harvest.

If the Japanese pot has a single message, it might be, "Be a passionate participant and not a cool observer. Involve yourself in every facet of your existence." For the potter, Japanese ceramics speak of a powerful and complete intimacy with the materials and processes of the craft. Mastering these fundamentals is tantamount to finding the art in them. For the user, Japanese ceramics display a diversity of types and an open, generous spirit of accommodation. This invites participation in the artistic process through assembling complementary pots, arranging their contents, and employing them imaginatively. Other arts, such as painting and sculpture, may shake the heavens, but life will always return people to utilitarian vessels.

In their quiet way, the ceramics of Japan offer a bridge between the everyday and the sublime.

DEFINING JAPANESE CERAMICS

Over the long history of ceramics in Japan, makers and users have developed ideas and attitudes about the craft that embrace both tangible things (materials, functions, techniques) and abstract ones (aesthetics, artisan identity). Five basic components might be identified.

First, Japanese ceramics show a diversity of materials and techniques. Despite Japan's comparatively small land mass and the apparent closeness of its population centers, the country's mountainous topography prevented extensive communication and homogenization. Many craft traditions developed in isolation. Villages were nearly closed socioeconomic systems, where craftspeople processed local materials to make objects for local populations. The materials of interest to the potters—clays, stones, pigments, and ashes—varied from region to region, and the properties of each influenced the means of manufacture and the appearance of products. Potters in one area might work in a coarse, white refractory clay that was congenial to ash glaze, while in the next valley workshops might utilize a smooth, red, fusible body that did not take glaze well but, after burnishing and firing at low temperature, yielded a rich crimson. Fortunately, there have long existed institutions—local daimyo sponsorship in the premodern era, and craft revival movements and tourism in the present century—to preserve these differences. The relationship between local materials and techniques and regional style can be observed in every pottery-producing culture, but nowhere has it been cultivated to such a high degree as in Japan.

Second, Japanese ceramics have developed a broad range of forms and functions. Even today, when many of the traditional kinds of ceramics have been replaced by objects made of metal and plastic, the variety is rich. One might take, for example, the category most familiar to Western functional potters: tableware. The contemporary Euro-American repertory is largely restricted to plates, bowls, and cups in a few standard sizes. Japanese food presentation typically makes use of numerous small- and medium-sized vessels rather than a few large ones. The range includes cups and bowls for liquids and soups, dishes of all sizes and shapes for food, cups and bottles for tea and saké service, as well as spice jars, chopstick rests, and even toothpick holders. Even in ordinary households, these are changed from day to day, depending on the food, the season, and the occasion.

In other categories, one can still see kitchen wares—storage jars, mortars, and pickling jars—in use in the countryside. Lidded earthenware cooking pots are used in restaurants. Wares for specific cultural and religious observances also continue to provide models for the potter. These include utensils for Buddhist and Shinto observances and for the tea ceremony and flower arranging. Although they are growing increasingly scarce, there are also handmade occupational wares such as net weights and octopus traps. Figure 2 shows the proportions and outlines of forms basic to Japanese ceramics.[1] Like vessel categories in the West, the nomenclature is derived from both function and shape. From left to right: the first two columns represent dishes and plates (*sara*), the second two columns large bowls (*hachi*) and small bowls (*wan*), the fifth column wide-necked jars (*kame*), the sixth column narrow-necked jars (*tsubo*), and the seventh column bottle forms (*bin*). There are, of course, numerous deviations: for example, when a certain bottle form is used for flower arrangement it is called a *bin*, but when the same vessel holds liquor it becomes a flask, or *tokkuri*.

Third, in Japanese ceramics the materials and techniques are seen as expressive in themselves. In Chinese ceramics lore, pots are praised for their resemblance to jade, silver, or ice. In Western ceramics, such as Italian majolica ware, every effort is made to recreate painting. But in Japan, the properties of the materials such as the color, texture, and hardness of the clay, impurity of the pigments, and viscosity and melt of the glaze are generally encouraged to speak for themselves. Potters are fond of saying that each material has an innate potential and that the job of the potter is to persuade that character to emerge. Bizen clay is rich in iron, and Bizen wares are ironlike. Kyoto clay has a fine, creamy white finish and thus is used for underglaze and overglaze decoration. Enabling materials to speak also influences form, which is generally left simple. In fact,

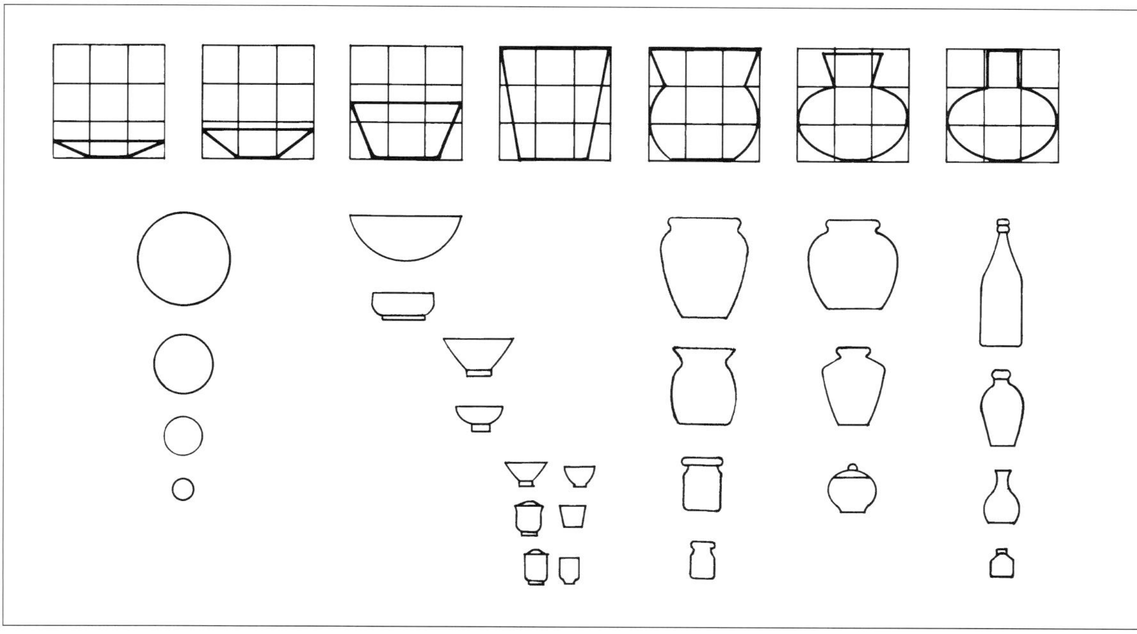

2. Basic forms of Japanese ceramics

symmetry, smoothness of contour, and complexity of decor—marks of sophistication in most ceramics-producing cultures—are not sought in many prominent Japanese wares.

The kiln is expressive as well. In most parts of the world, the kiln is seen as a device to harden the clay and melt the glaze. This is also true in Japan, but for at least four centuries, expressive kiln effects such as warped shapes, scorch marks, and irregular ash deposits have been pursued deliberately. In other words, the kiln is an active agent in the process of decoration. Few potters would say, however, that the caress of the kiln is simply a divine gift. All art contains artifice, and Japanese ceramics are no exception. Potters who excel in "making the kiln speak" are able to do so only after arduous trial and error. For example, a great deal of planning goes into loading and firing a kiln for unglazed stonewares, and good potters readily know the difference between a pot with a fine accidental ash "landscape" and one that is simply overfired. These fortuitous effects do not result from simply letting it happen, although some Western observers seem to think so. Nature is certainly more powerful than any human enterprise, but most potters reach that conclusion only after a hard course of training. One purpose of this book is to locate the place of serendipity and naturalness on one hand, and deliberate, diligent effort on the other. In Japanese ceramics, the balance comes through mastering and transcending technique, not rejecting it outright.

Fourth, ceramics enjoy a level of appreciation in Japan unmatched in any other culture. Most people, the Japanese included, use ceramics without a continuous, conscious aesthetic appreciation. But there is a world of difference between unselfconsciousness and plain insensitivity. While most Japanese people do not stop to evaluate every pot they encounter, many feel a vague sense of uneasiness if there is something strange about the form, color, or weight of a vessel. Part of this is prompted by the use of smaller vessels that are frequently picked up and set down in the course of a meal. But there is also the issue of the objects themselves, the materials and designs. Americans in particular, raised with industrial china or, for that matter, paper plates and plastic foam cups, have little common sense about how a good pot should look or feel. How can one begin to evaluate and appreciate functional ceramics in the absence of such a basic measure? All the seductive philosophies and heady abstractions used in treatises on Japanese ceramics—such as Zen, tea, and folk art—will never replace the simple but deep realizations that come from using good pots every day. In premodern Japan, this appreciation for the basics of craft ran deep and wide. When the pioneering Japanologist Edward Morse toured the islands in 1878, he remarked how country folk would gather around him, eagerly touching his then-exotic woolen garments and speculating about how they were made.[2] It is from this level of curiosity and esteem that crafts of quality are made.

Behind this simple appreciation are several distinctive cultural factors. First, the Japanese people tend to revere things that are simple and unadulterated. Regardless of whether this outlook is a remnant of Japan's ancient animistic traditions of worshiping natural objects, or a more measured reaction to the complexities of foreign cultures, such an attitude has certainly encouraged craftspeople to evoke a straightforward, natural beauty in their work. Second, there is an interest in making these simple things work in expressive combinations. This has been particularly nurtured in the tea ceremony (fig. 3). Beginning in the sixteenth century, tea masters began to advocate mixed groups of utensils rather than sets; they saw irregular groupings as having an appealing degree of incompleteness, even a poetic sadness. This can be likened to how a Western person might feel when browsing in an attic full of old, randomly preserved

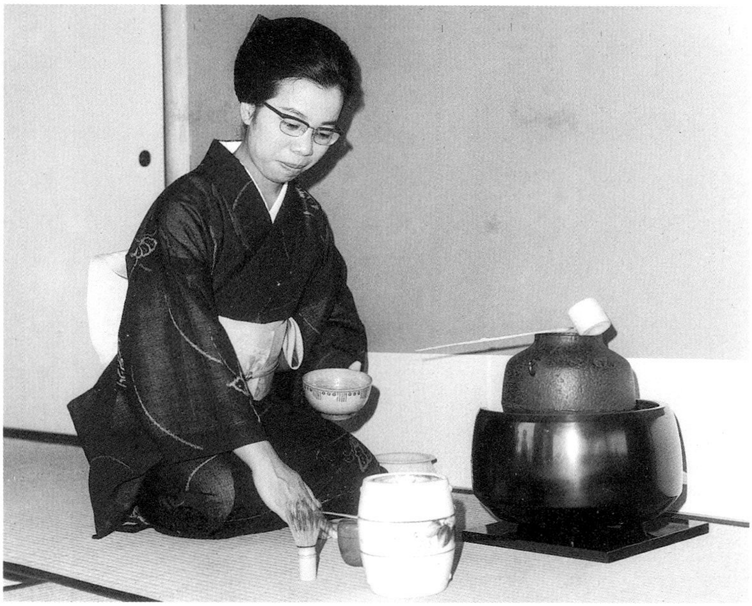

3. Tea ceremony

belongings. The contrasts in such mixed groups of vessels also enhanced the formal qualities of any single object. It has become a convention to employ different shapes, textures, and colors in a single ceremony. Guests admire not only the shape and decoration of a pot, but the texture of the clay and glaze, the way the object feels in the hand, and even how the surfaces mature with age.

And then there is handling. Handling a teabowl, for example, means not only drinking from it but examining all of its parts, especially its base. On many Japanese pots, a foot ring is carved out of the base when the vessel is half-dry. Since it is usually unglazed, it is where the potter's manual agility and sense of proportion are most visible. Even though it is an unseen part of the pot, the foot ring is in fact a focal point for assessments of merit. Bold or hesitant, generous or niggling, crisp or dull, spontaneous or calculated—such is the range of values attached to this humble domain. The extent of this appreciation, perhaps more acute when it is denied, is evident in this account by collector Kuroda Ryōji:

> Around 1965, the Shino ware teabowl *Mine no Momiji* (Summit Maple Leaves) was put on display. We were not allowed to touch the bowl, which was set on a purple cloth, but we could see it from all four sides. The restrictions placed on the viewing were cruel indeed, for to feel that bowl resting in the palm of one's hand surely would have been a more meaningful encounter with Truth. It was like having an itch one could not scratch.[3]

Seeing, which we all too often equate with knowing, is insufficient to grasp the messages embedded in these pots. Because the maker knows that the pots will be used, they are traditionally left "incomplete." Because they are awaiting completion by the user, the shape and decoration do not in themselves overpower the senses. Appreciation and use are intertwined. Japanese intimacy with nature, given specific form in the tea ceremony, has worked against the division of the sacred and secular so common to the utensils of developed societies. What an advantage for the functional potter!

A fifth and final point is that the Japanese potter is expected to develop spirit along with skill. Just as users come to see and expect transcendental qualities in their pots, potters become mindful of the need to possess something beyond technique. At least as early as the fourteenth century, one can see traces of the belief that artistic mastery is a spiritual way found in the depths of a person's inner experience rather than the acquisition of technique or the display of talent. A Zen Buddhist expression states it thus: "To obtain a certain thing, you have to become a certain person." Traditional pottery apprenticeship stresses incidental tasks, silent observation, and trial and error. The teacher does not tell the student how to position the fingers in making a rim, how long to grind a pigment, or when to stoke the kiln. This is a lonely road, but the solitude provides focus and self-reliance. Learning each task from scratch through both body and mind instead of tackling information through the mind alone makes one cognizant of internal principles and their possible applications. This takes longer than learning things in a classroom, but it provides a long-term framework for creativity and growth. Users of pots enjoy watching the course of that journey in their favorite potters. The joint venture of mindful potters and enlightened consumers has contributed much toward the recognition of ceramics as a fine art in Japan.

A SHORT HISTORY

Whether it attempts to follow the past, ignore it, or actively repudiate it, a potter's work is always related to history. The Japanese potter has traditionally sought to evoke the past, either by modestly following an inherited workshop style or by more consciously blending

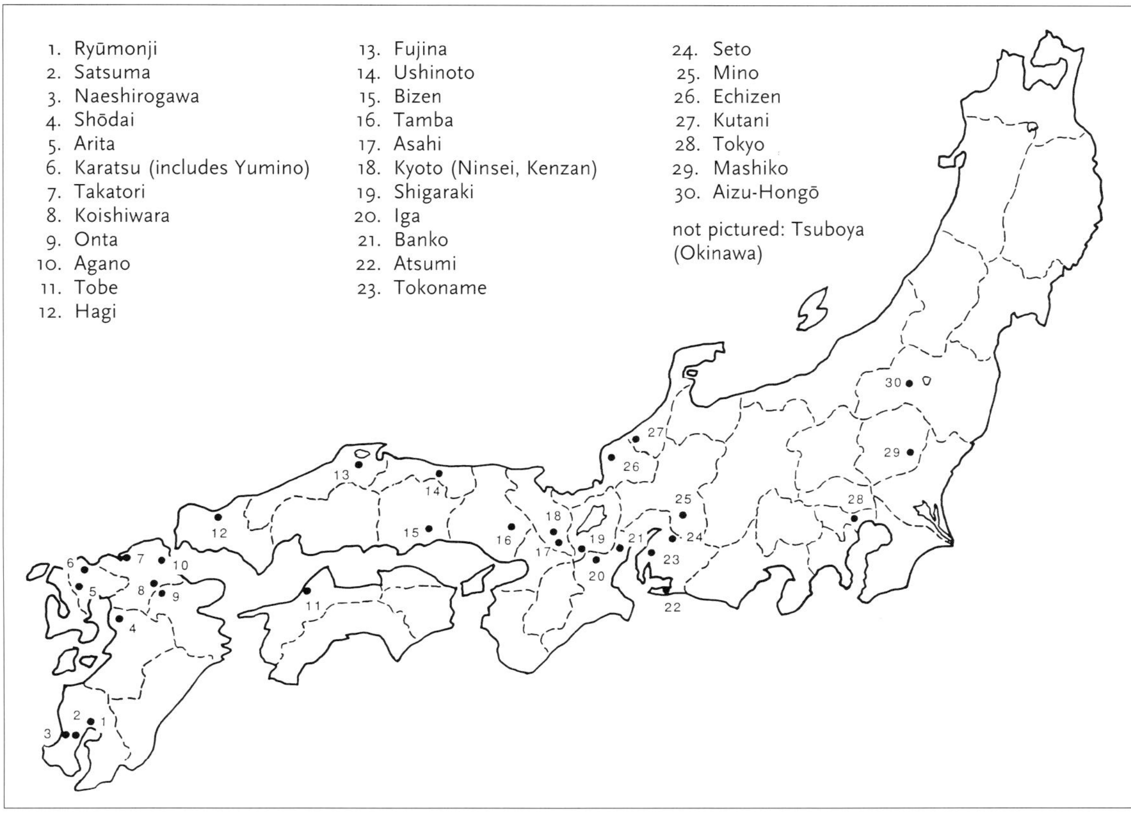

4. Map of kilns mentioned in text

elements from the national heritage. This section surveys the fund of techniques that potters developed at kilns nationwide (fig. 4) and shows how that fund has served later generations.

Early Earthenwares

Recent archaeology suggests that ceramics were first made in Japan about ten thousand years ago. These earthenwares, called Jōmon ("cord-marked") after their surface patterns, were built from coils or slabs and were finished by smoothing or perhaps paddling the joints between coils (fig. 5). With the exception of the very earliest deep bowl forms, the common early form was a pointed-bottom vessel. A flat-bottomed cylindrical shape, sometimes featuring an individualistic, fantastically sculpted superstructure, later replaced the pointed type. The wares were pit-fired in a partially reduced atmosphere to a temperature of about 700–750 degree centigrade. There seems to have been little influence from mainland Asia until the end of the Jōmon period. From about 300 B.C.E., however, a different type of earthenware emerged, one related to newly arrived rice cultivators. The style is called Yayoi, after the area in Tokyo where it was first

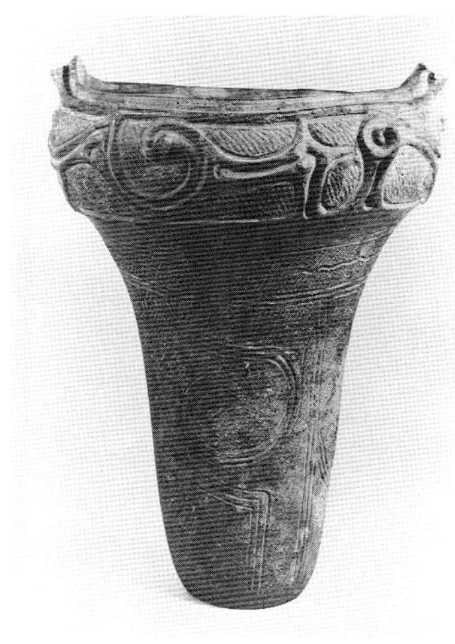

5. Jōmon ware

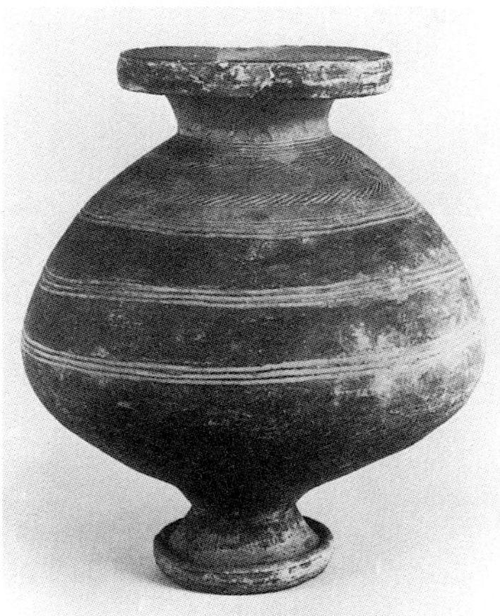

6. Yayoi ware

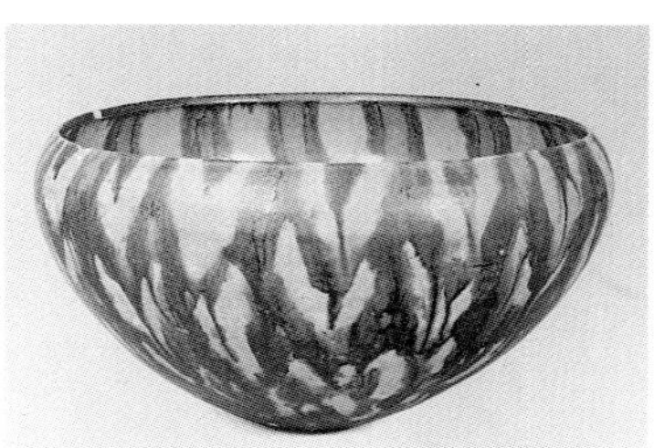

7. Three-color ware

discovered. Like the Jōmon earthwares, Yayoi wares were coil-built, but with thinner walls, more regular contours, and often with smooth painted or burnished surfaces instead of plastic decor (fig. 6). Yayoi forms were sometimes finished, although not fabricated, on a primitive turntable. No Yayoi kilns have been discovered to date, but the firing is thought to have taken place in the open, with straw and hardwood twigs as fuel. The reddish surfaces of Yayoi ware show a comparatively clean-combusting or oxidizing atmosphere, as opposed to the brownish, partially reduced Jōmon wares.

Around the third century of the common era, Yayoi wares were replaced by a ware called Haji, a term used to designate earthenware in Japan's earliest written histories. Like the Yayoi wares, the Haji wares were coil-made with occasional finishing on a crude turntable. Haji potters also pioneered the forming of wares from a single slab. They fired their wares in shallow pits, some of which have been discovered. The comparatively crude appearance of the product is attributed to mass production, and that is corroborated in documentary references to *hajibe,* entire families or guilds specializing in the manufacture of ceramics. The *hajibe* also appear to have been charged with the manufacture of *haniwa,* the earthenware effigies found in the large tomb precincts of the fifth to seventh centuries.

The first appearance in Japan of applied glazes occurs in the so-called three-color ware (there are also single-color and two-color types), a type of earthenware that took inspiration from the polychrome sepulchral wares of China. The Japanese versions were produced in the Nara area from the early eighth century (fig. 7). These wares were also the first double-fired Japanese ceramics: the raw ware was hardened in a bisque firing, and a lead-glaze coating melted in a subsequent glaze firing. The Shōsōin Imperial Repository in Nara has preserved a record dated 733–34 that lists ingredients for making polychrome lead glazes. The key ingredients, metallic and lead oxides and siliceous rock powder, are basically the same as those used on all subsequent Japanese glazed earthenwares. The kilns, probably simple cylindrical chambers fired from beneath, were the earliest low-temperature kilns constructed in Japan. In the ninth century, potters also began to use the lead-glaze technology to imitate celadons, specifically the popular Yue celadon wares of southeast China. These so-called green-glazed wares were first produced for ceremonial purposes, but by the tenth century they had become decidedly utilitarian.

The Emergence of Stoneware

From the late fourth or early fifth century, a new style of ware with Korean origins began to be used in Japanese burial offerings. This Sue ware (fig. 8) was characterized by several key innovations, including the deliberate selection of refractory clays, use of the potter's wheel, and firing in an excavated tunnel kiln capable of achieving the low end of the stoneware spectrum, about 1100–1200 degrees centigrade. An excess of fuel was introduced at the very end of the firing, which turned the surface of the pots a charred gray. An early and extensive kiln complex, in continuous use from the fifth through eleventh centuries, was located in Suemura, near the present-day metropolis of Osaka. The recent discovery of very early Sue kilns in many other areas, however, seems to discount an old theory that the style diffused from a central location in Japan. But in any case, Sue-ware sites multiplied throughout the country and became one root of the medieval stoneware tradition that followed. The early Sue vessels followed the ceremonial prototypes from Korea, but plates, bowls, and ewers dominated production from the seventh century. Sue-ware food vessels also widely began to replace their wooden counterparts, and since they were waterproof, Sue jars proved advantageous for storage. On the other hand, the low resistance of these stonewares to thermal shock also guaranteed that the soft but refractory Haji earthenwares would continue to find use as stoves, steamers, and cooking pots until iron wares began to replace them much later, in the Kamakura period (1185–1333). The ability of such conservative technologies to survive alongside more innovative ones is a distinctive aspect of Japanese ceramics.

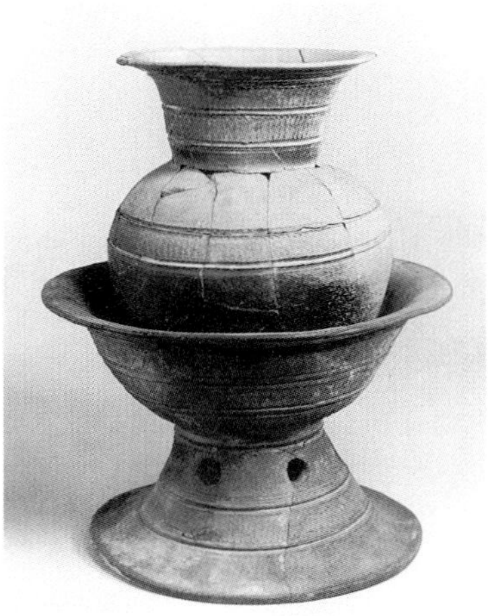

8. Sue ware

Around the ninth century, a Sue-ware center at Sanage, Aichi Prefecture, began a new type of high-temperature work. Here, too, Yue-ware celadons imported from China provided the chief inspiration. The Japanese products, now known as ash-glazed wares (fig. 9), were deliberately coated with a suspension of wood ash that melted into a glossy but somewhat fluid glaze. As with Sue wares,

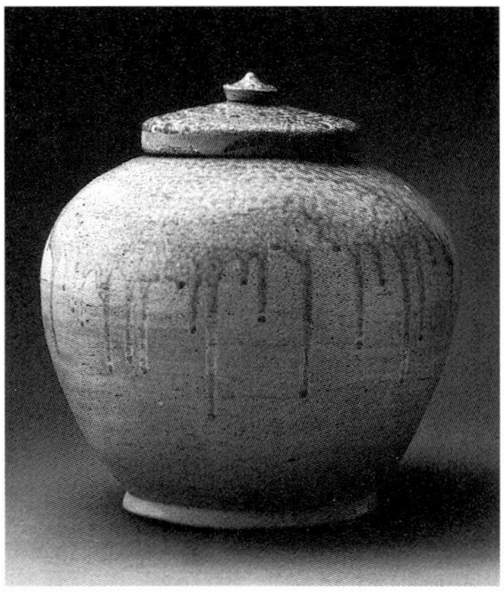

9. Ash-glazed ware

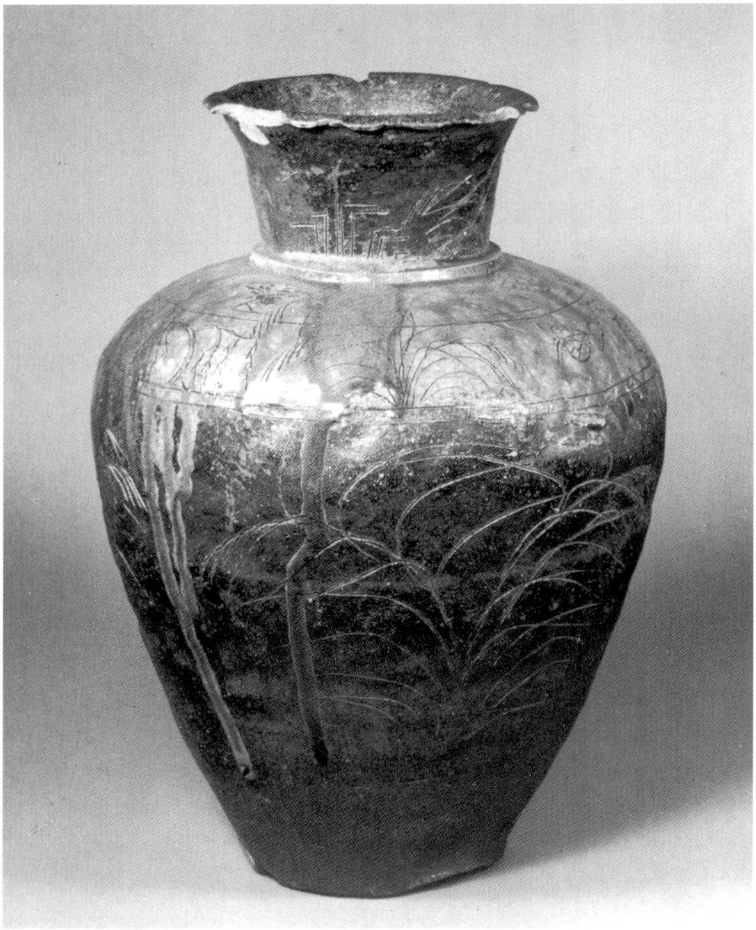

10. Atsumi ware

ordinary plates, bowls, and cups gradually augmented the initial output of ritual products. From around the tenth century, the Sanage potters installed a dividing pillar in the middle of the kiln with the apparent purpose of increasing kiln capacity and improving heat distribution. The design, which had no precedent on the Asian mainland, seems to have reduced the number of kiln losses. Other workshops in the Sanage area, seeing the benefits of increased production, began to adopt the Sanage techniques. Prominent among the followers were the Tokoname (see pl. 1) and Atsumi (fig. 10) kiln centers, which would become important producers and technique disseminators in the age that followed.

Due to internal disturbances in China, whose exports served as models for many Japanese products, and to the decline of the urban aristocracy at home, glazed ceramics ceased to be made in Japan in the eleventh century. On the other hand, expanded agricultural production led to mass consumption of unglazed ceramics throughout the country. Sanage-affiliated centers like Tokoname and Atsumi began to produce large amounts of pottery, particularly wide-mouthed jars, narrow-mouthed jars, and kitchen mortars. The traditional coiling technique was employed in making these wares. When the wheel was used, it was only for starting and finishing.

Styles and techniques from Tokoname and Atsumi began to spread all over the country, rivaling the older Sue traditions. Over thirty kiln centers of this medieval period, which lasted from the twelfth through the sixteenth century, have now been identified. But only those with flexible marketing strategies and attractive products managed to survive into later times. These include Echizen, Tokoname, Shigaraki, Tamba, and Bizen, all of which produced distinctive unglazed stonewares, and a single region that produced glazed stonewares, Seto-Mino.

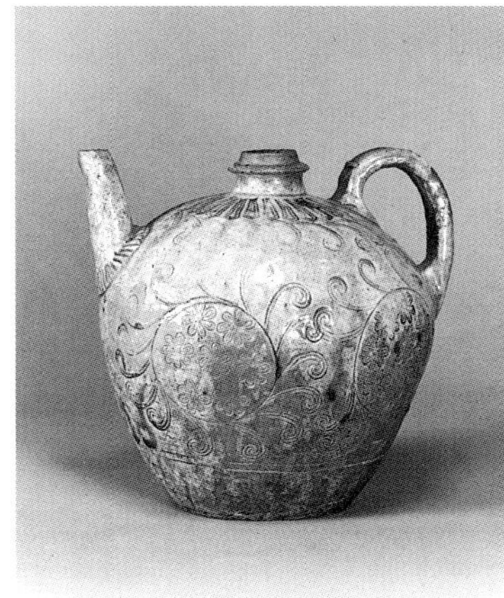

11. Seto ware

Seto, still the largest pottery-producing region in Japan, was an old Sue center that fell under Sanage influence at the beginning of the twelfth century. About a century later, renewed contacts with China once again made purchase of glazed ceramics a possibility for the highest elite. Lower-ranking aristocrats purchased Seto wares, which took after the imports (fig. 11). Early products from Seto featured simple ash glazes with little improvement over the old Sanage types. From the fourteenth century, however, glazed-ware production at Seto expanded dramatically. The wheel was restored as the exclusive forming tool, the yellowish-green ash glaze was stabilized, and an opaque iron glaze (*temmoku*) was developed. These techniques also spread to nearby Mino. Products included not only large jars, ewers, and bottles, but also small containers for powdered tea and bowls for drinking the beverage. It is at this time that tea drinking began to have an impact on Japanese ceramics, becoming a powerful factor behind the emergence of a truly native stoneware.

Maturation of the Native Style

Tea was introduced to Japan from China in the eighth century, but the development of the tea ceremony actually started about five hundred years later, when the Zen monk Eisai (1141–1215) introduced to Japanese monastic life the Chinese way of whisking powdered tea in hot water. The brew was considered efficacious as a medicine and as a stimulant during meditation. Within a century, tea also came to be enjoyed as part of extravagant warrior-class social gatherings. Exotic Chinese utensils were not only used, but exhibited and even vied for in elaborate contests. As the warriors' utensil collections grew, personnel were recruited to oversee acquisitions and to serve tea. When the nation was engulfed in civil war during the second half of the fifteenth century, however, the mood of some tea ceremonies changed. To be sure, proud possession continued unabated, but a sense of restraint now marked the tea-drinking event.

Prominent members of the merchant class, which had come to enjoy considerable autonomy in the age of turmoil, experimented with the tea-ceremony environment and its utensils. Several merchant tea men, including the celebrated Sen Rikyū (1522–91), were chosen to serve the flamboyant warlords Oda Nobunaga (1534–82) and Toyotomi Hideyoshi (1536–98). A new aesthetic for Japanese ceramics developed in that environment. As previously mentioned, the earliest utensils were imported from China, and the continental standard guided the first production of tea wares at Seto. By the early sixteenth century, tea men had begun to see in rural Japanese ceramics (unglazed stonewares of the medieval tradition) a stark, severe beauty quite unlike that of the Chinese models. Zen Buddhist exhortations to humility and the aestheticized poverty evoked in native literature inspired tea men to appropriate artless objects such as seed jars and hemp buckets for use in tea service. By the end of the century, tea masters like Rikyū were consciously designing ceramics, especially a kind called Raku ware, in this mode.

Besides its significance for the tea ceremony, Raku (see pl. 12; fig. 12) represents a renewed level of enthusiasm for earthenware. It is interesting that, despite the development of stoneware and later porcelain, the earthenware tradition never expired in Japan. Earlier it was noted that the unglazed Haji earthenware production continued along with that of high-fired Sue stoneware. In Kyoto, green-glazed earthenware and tile were manufactured for elite consumption in the ninth and tenth centuries. Although these glazing techniques disappeared in the eleventh century, unglazed earthenware continued to be made. The return of low-temperature glazing is evidenced in the appearance of Raku wares. The small corpus of early

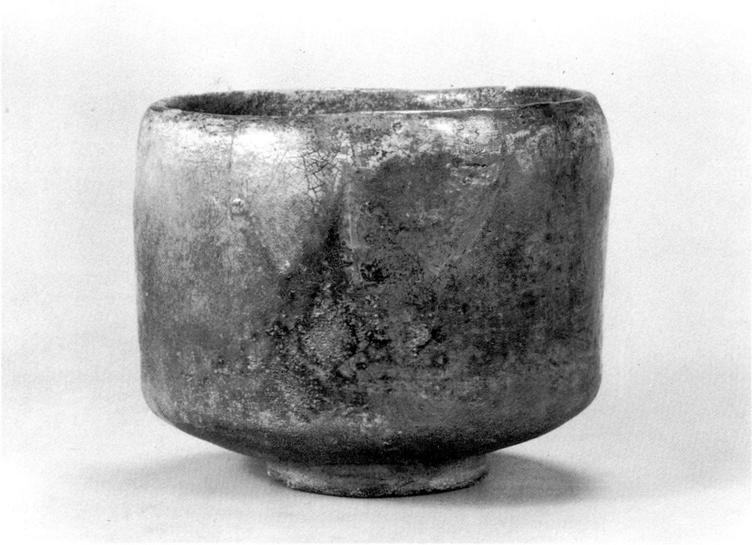

12. Raku ware

Raku-ware works, mostly tea bowls, is thought to be the result of collaboration between tea master Rikyū and a Kyoto artisan named Chōjirō (d. 1589). The objects are handmade, coated with black and transparent lead glazes, and fired individually or in small numbers in a small updraft kiln. Because of its artless shape, superb insulating qualities, and ease in handling, the Raku-ware tea bowl became the preferred vessel among tea masters.

If Kyoto was the site for tea-ware design and patronage, the locomotives of production were Mino, a region to the east of the capital, and Karatsu, in faraway Kyushu. In the fifteenth century, the Mino area, which possessed rich supplies of feldspar and refractory clay, began to surge ahead of neighboring Seto in the struggle to emulate Chinese Ming-dynasty ceramics. When orders went out for innovative tea wares a century later, Mino had the personnel and technology, including a new kind of kiln built largely above ground, to accommodate the demand. The great Mino tea-ware styles—Yellow Seto, Black Seto, and Shino wares—were based on the older Seto glazes, but with dramatic modifications in shape, decoration, and finish. The glaze of Yellow Seto (see fig. 105), the oldest of these styles, combined the Seto ash glaze with clay, perhaps in an attempt to produce celadon, but a devitrified yellow surface resulted instead. Black Seto (see fig. 131), which is believed to have been produced at the same time as its Kyoto counterpart, Raku ware, made use of the traditional Seto iron brown glaze, but the sudden withdrawal of the product (mostly teabowls) from the kiln at peak temperature froze the color to a stony black. Shino ware (see pls. 7-9) seems to have

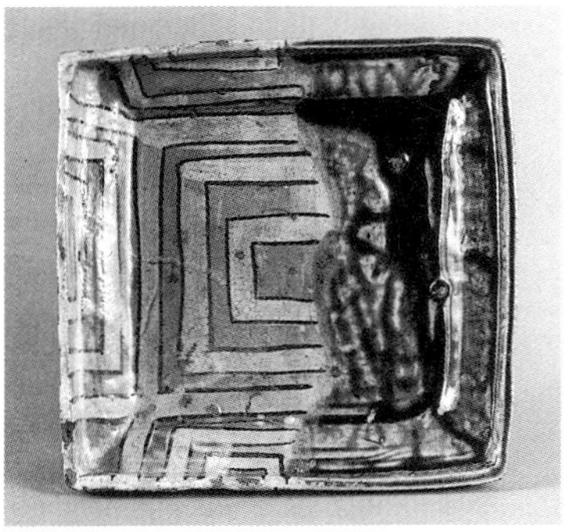

13. Oribe ware

resulted from the search for white ceramics with underglaze painting, but the use of a glaze high in feldspar resulted in its distinctive snowy texture. From about 1600, the assertive taste of tea master Furuta Oribe (1544–1615) appears at Mino. The unusual shapes of Black Seto bowls, the copper-green accents on Yellow Seto, and the transparent Shino glaze all combined in the diverting and idiosyncratic Oribe wares, which gained special popularity as food containers (see pls. 10, 11; fig. 13). Production was streamlined by the introduction of the stepped domical climbing kiln at Motoyashiki around 1600. At the same time, other regional kiln products, both glazed and unglazed, began to exhibit the same interest in deformation.

Kyushu kilns, especially the centers of Karatsu and Arita, were all-important sites for technology transfer from the continent in the sixteenth and seventeenth centuries. Trade contacts with Korea and, later, two invasions of the peninsula (the so-called pottery wars of 1592 and 1597) stimulated the production of Karatsu wares (see pl. 4). Those wares introduced to Japan the fast-turning kick-wheel, transparent and opaque feldspathic glazes, and the stepped single-vaulted climbing kiln, soon to be followed by the classical stepped domical-chamber climbing kiln, which was probably introduced to Kyushu from China and in turn transmitted to Mino. The Korean-inspired techniques and styles spread throughout Kyushu and western Honshu, inspiring traditions such as Agano, Takatori (see pl. 5), Satsuma (see pl. 22), and Hagi (see pl. 6). Feudal lords who were attracted to tea-ceremony wares and who had an interest in developing local industry sponsored these kilns.

Porcelain and Decorated Stoneware

Korean potters are also believed to have been responsible for the discovery of deposits of porcelain stone in Arita, and by the second decade of the 1600s porcelain was being produced at kilns in the area. Prominent among the early wares at Arita was decoration in underglaze cobalt, or blue and white (see pl. 16; fig. 14). In the 1640s the overglaze enamel technique—pigmented glasses fused to the glazed surface (usually porcelain) in an additional firing—was added to the repertory. Large-scale exports to Southeast Asia and Europe commenced within a few years. Although the initial painted

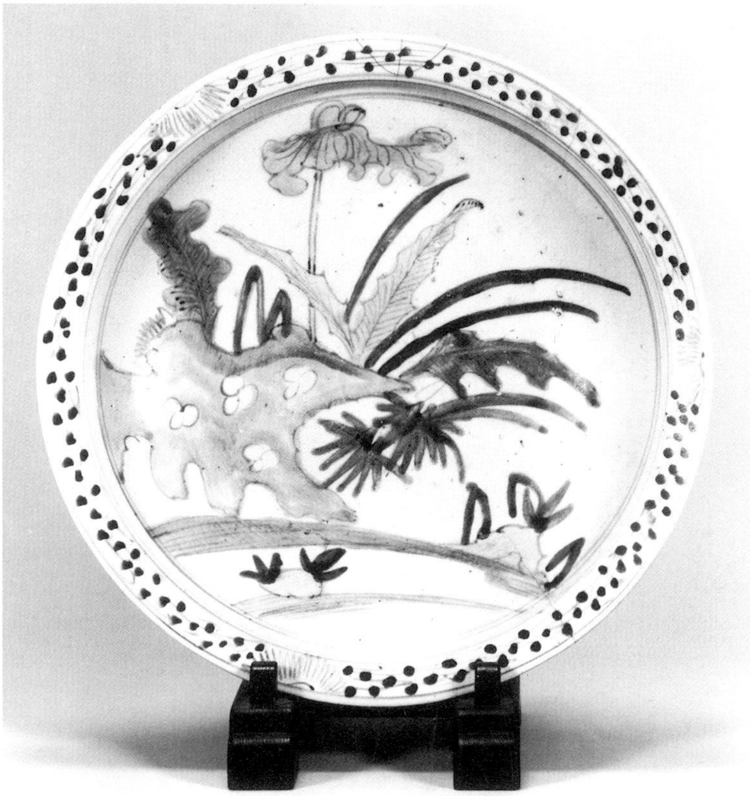

14. Arita ware

designs followed Chinese and Korean precedents, Japanese shapes and design schemes became popular in the second half of the seventeenth century. Recent excavations and scientific tests also suggest that early Kutani ware, traditionally thought to have been developed in Ishikawa Prefecture in the 1650s, was actually part of Arita's domestic line (see pl. 18). In any case, Arita's focus on fine cuisine wares, on painted decoration, and on year-round production by teams of specialists represents the beginning of a large-scale ceramic industry.

Besides Seto-Mino and Kyushu, the other major ceramic center in early modern Japan was Kyoto, the imperial and cultural capital of the nation. Large-scale stoneware production commenced there in the early decades of the seventeenth century, probably by skilled Seto artisans lured by the burgeoning interest in the tea ceremony. In the mid-1600s, an initial interest in subdued and elegant tea-ware styles began to give way to the manufacture of decorated ceramics. The greatest resource for the Kyoto ceramics industry has always been its proximity to centers of patronage. Thus it is not

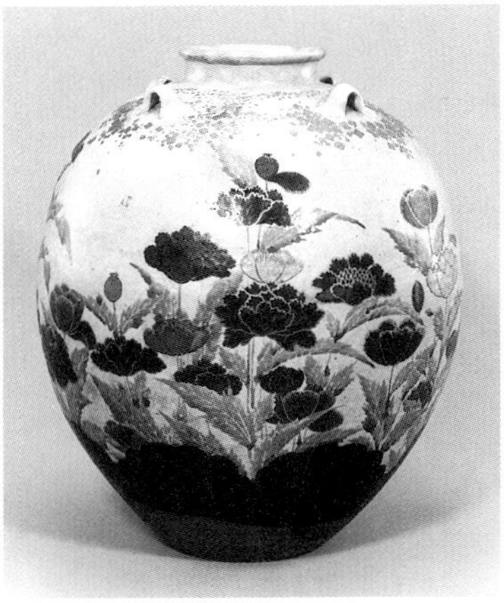

15. Ninsei ware

surprising to see in Kyoto examples of close collaboration between patrons and artisans, and in some instances potters who themselves were highly cultivated in the arts and letters. An example of the former is Nonomura Ninsei (active second half of 17th c.; fig. 15), who was advised by experts in the tea ceremony, and among the latter is Ogata Kenzan (1663–1743; see pls. 13, 14), originally a dilettante who pursued aristocratic pastimes such as poetry and calligraphy. Both of these men were instrumental in bringing the sophisticated brushwork of painting and calligraphy into the potter's art. Major centers like Kyoto also became sources of dissemination to outlying kilns, which had begun to grow in number in the eighteenth century. Warrior and merchant patrons from the provinces sent their artisans to Kyoto for study with tea masters or master potters and invited Kyoto potters to visit them. The flow of information also encouraged the compilation of potter's manuals, which began to circulate in large numbers from the early nineteenth century.

Contact with the West

The rapid modernization of the Meiji era (1868–1912) had a profound impact on Japanese ceramics. The presentation of official gifts and the tea ceremony, which provided the highest earnings for potters, all but disappeared. Japanese potters looked to export as the future of the industry, and by the mid-1870s they were winning recognition at international expositions in Europe and the United States. Exchange with the West also facilitated access to techniques such as slip-casting, new varieties of overglaze enamels, and fossil-fuel kilns. Mass production and the export drive naturally took a toll on the older traditions, a fact that came to the attention of both critics and younger potters in search of alternatives. In the early twentieth century, two movements arose that did much to revitalize production of smaller workshops: the folk art movement and the tea-ceremony revival.

The Japanese folk-art *(mingei)* movement was inspired by Western design reformers like John Ruskin and William Morris as well as by the tea-ceremony tradition of elevating vernacular objects to the status of works of art. The *mingei* movement was officially inaugurated in 1926 by its leader and spokesman Yanagi Sōetsu (1889–1961), a university-educated philosopher who believed that

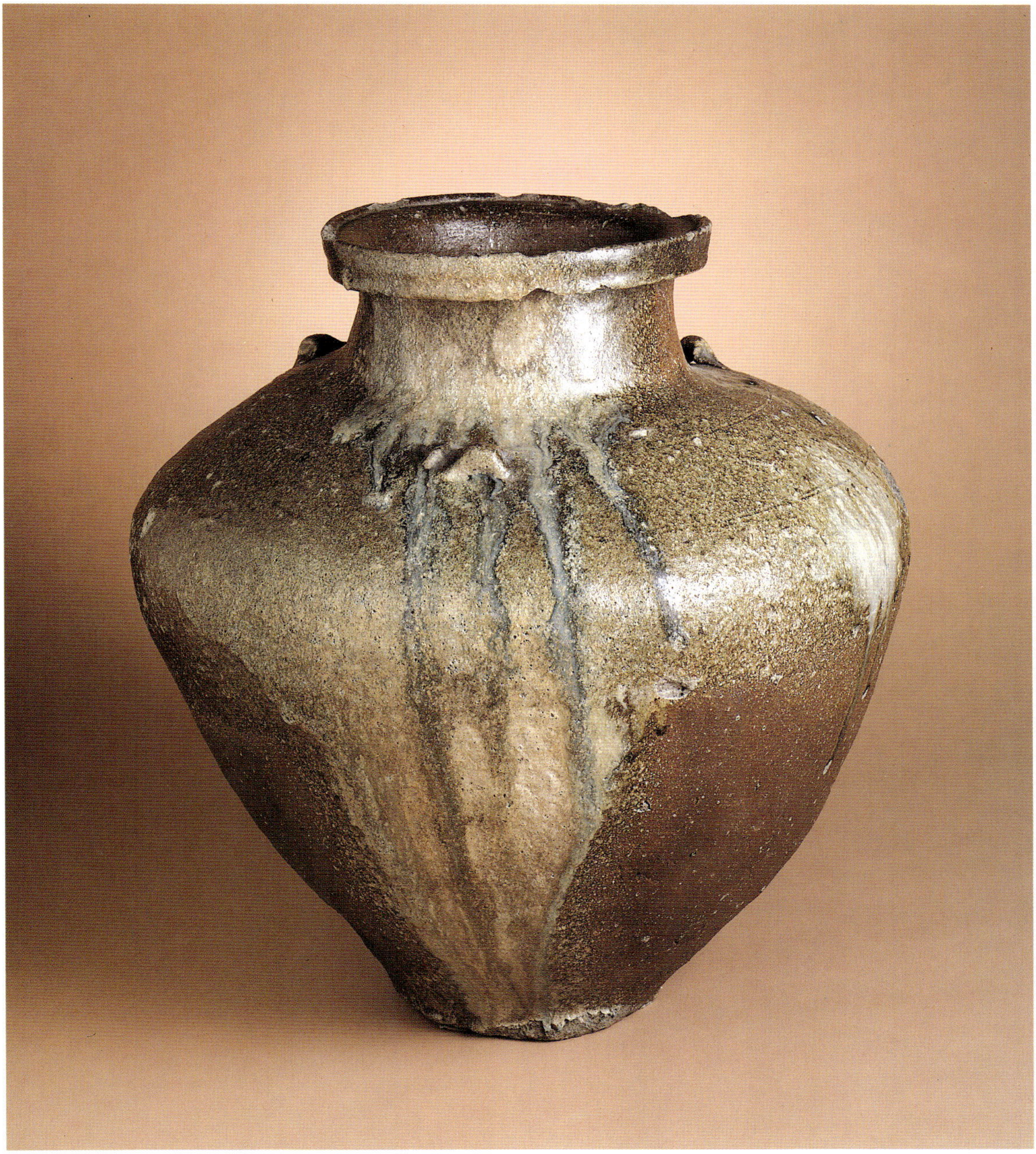

1. *Tokoname ware: Storage jar with natural ash glaze. 14th c. Stoneware, ht. 45.0 cm., diam. 45.3 cm. Los Angeles County Museum of Art.*

Tokoname ware, produced on the Chita Peninsula in present-day Aichi Prefecture, was one of several dozen kiln centers that arose in the twelfth and thirteenth centuries in response to increased agricultural production. Its large storage jars are particularly distinctive. The clay is minimally processed, with large grains of quartz remaining. These protrude from the surface in the firing, imparting a coarse texture. The glaze is accidental, resulting from fuel ash that has settled on the pots during firing in an excavated "tunnel" kiln.

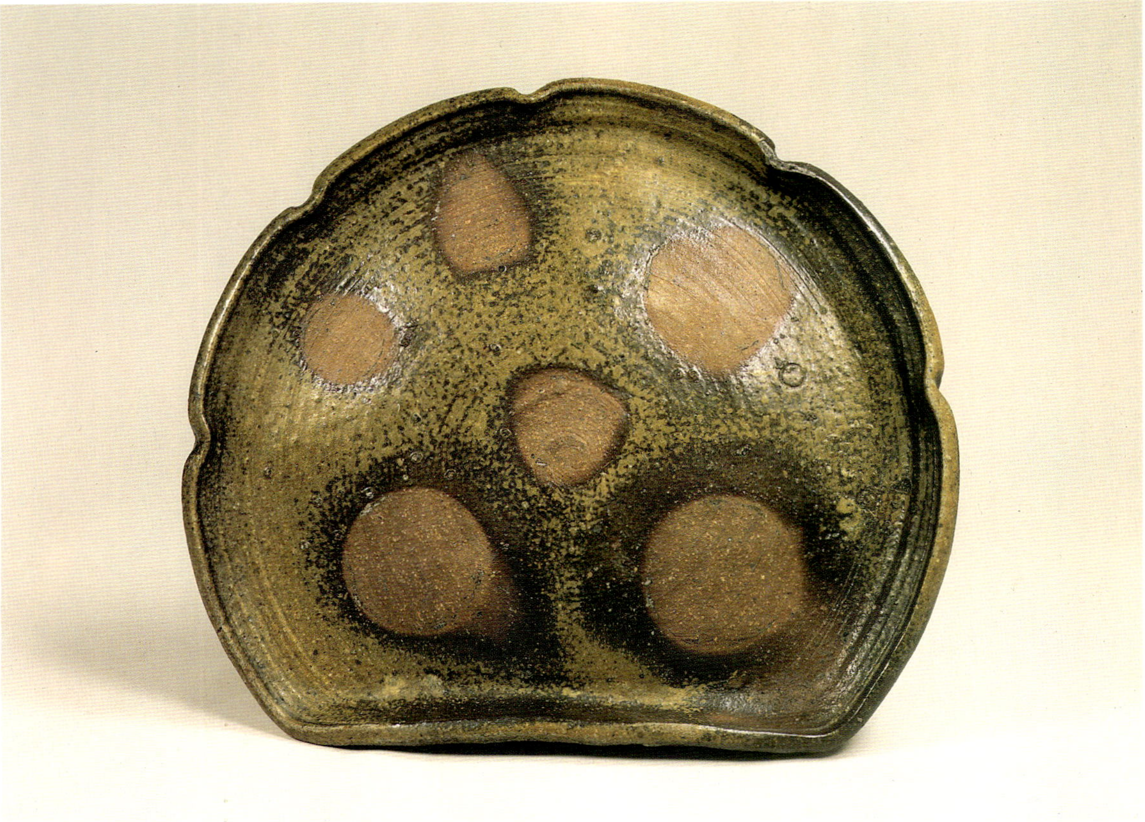

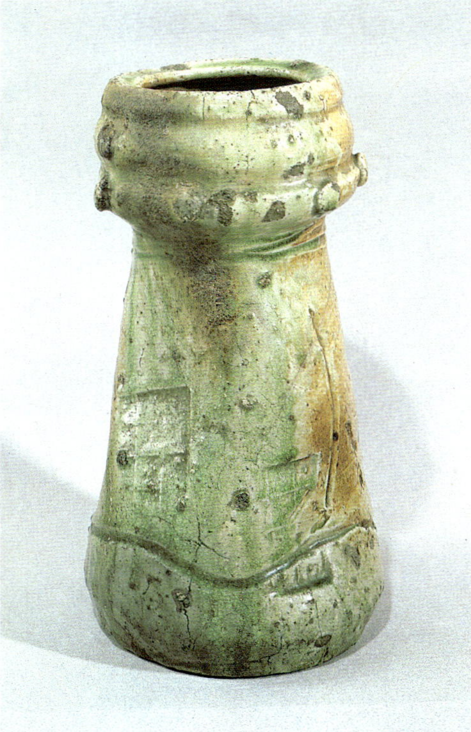

2. *Bizen ware: Serving dish with resist patterns. Early 17th c. Stoneware, ht. 4.1 cm., diam. 33.8 cm. Tokyo National Museum.*

Bizen ware, an unglazed stoneware produced from the thirteenth century in present-day Okayama Prefecture, traditionally was stacked in the kiln without the aid of saggars or shelves. When round objects such as bottles or small dishes were stacked on flatware, the shielded areas would be protected from open flame and ash. From the early seventeenth century, potters exploited this effect as decoration, calling it *bota-mochi*, after a popular disc-shaped sweet.

3. *Iga ware: Flower vase with paddled and carved designs. Early 17th c. Stoneware, ht. 28.8 cm. Private collection, Japan.*

The shape of this vase is derived from the metal flower containers imported from China for use in Buddhist ceremonies in Japan. From the late sixteenth century, the Iga potters, centered in what is now Mie Prefecture, began to freely interpret such shapes, subjecting them to multiple high-temperature firings. Three principal effects include reddish scorch marks, deposits of a greenish natural ash glaze, and the fissured "stone explosions" caused by large fragments of quartz in the clay.

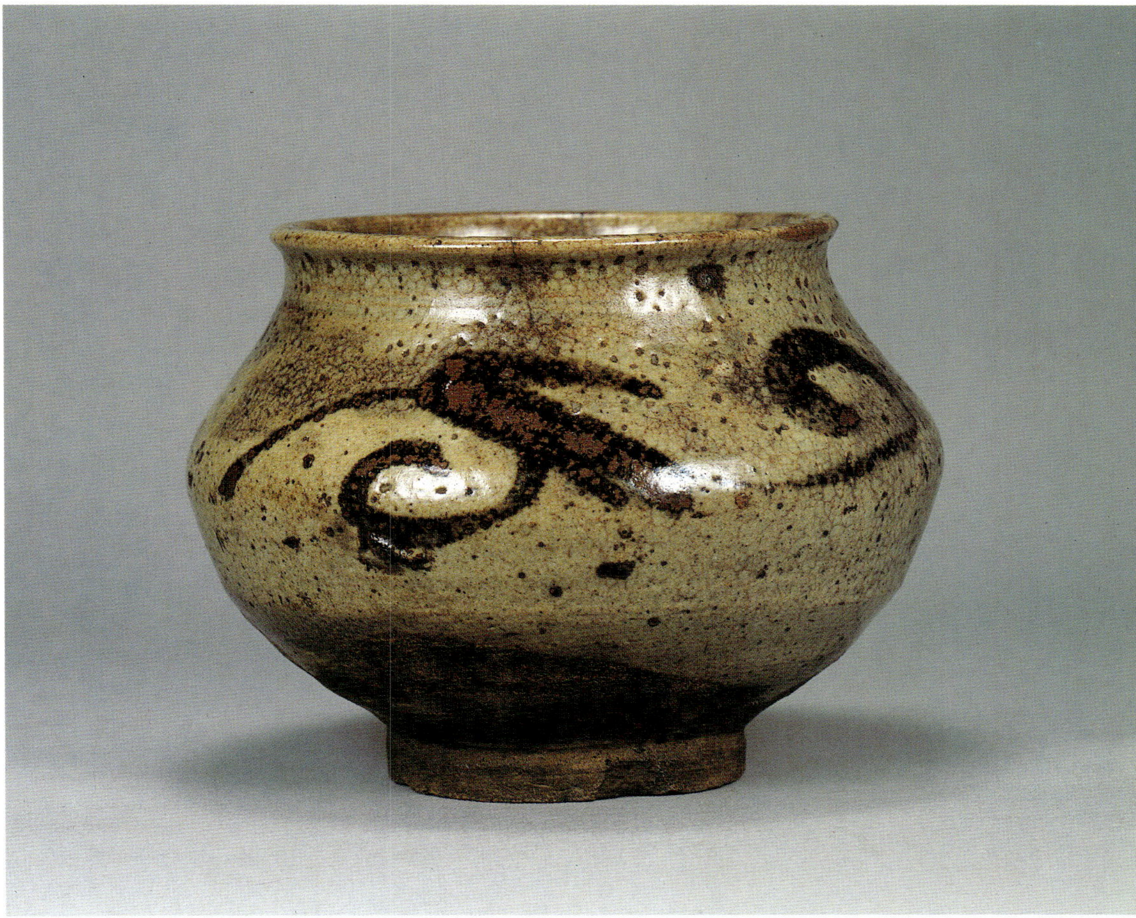

4. Karatsu ware: Water jar with abstract design in underglaze iron. Early 17th c. Stoneware, ht. 11.9 cm., diam. 12.0 cm. Private collection, Japan.

Karatsu ware was produced from the late sixteenth century at kilns located in Saga and Nagasaki prefectures on the island of Kyushu. Many of the early potters were from Korea, and the shape of this jar strongly suggests a Korean prototype. After a rapid forming on the kick wheel, a design was freely brushed in iron pigment. The iron melted into solution when the ash glaze was fired to stoneware temperature, but during cooling it precipitated to the surface in metallic crystals, creating the "rust" effect. Like most Karatsu wares, this jar was probably made for ordinary kitchen use, but its unassuming beauty was "discovered" by tea ceremony masters, who used it as a fresh water container.

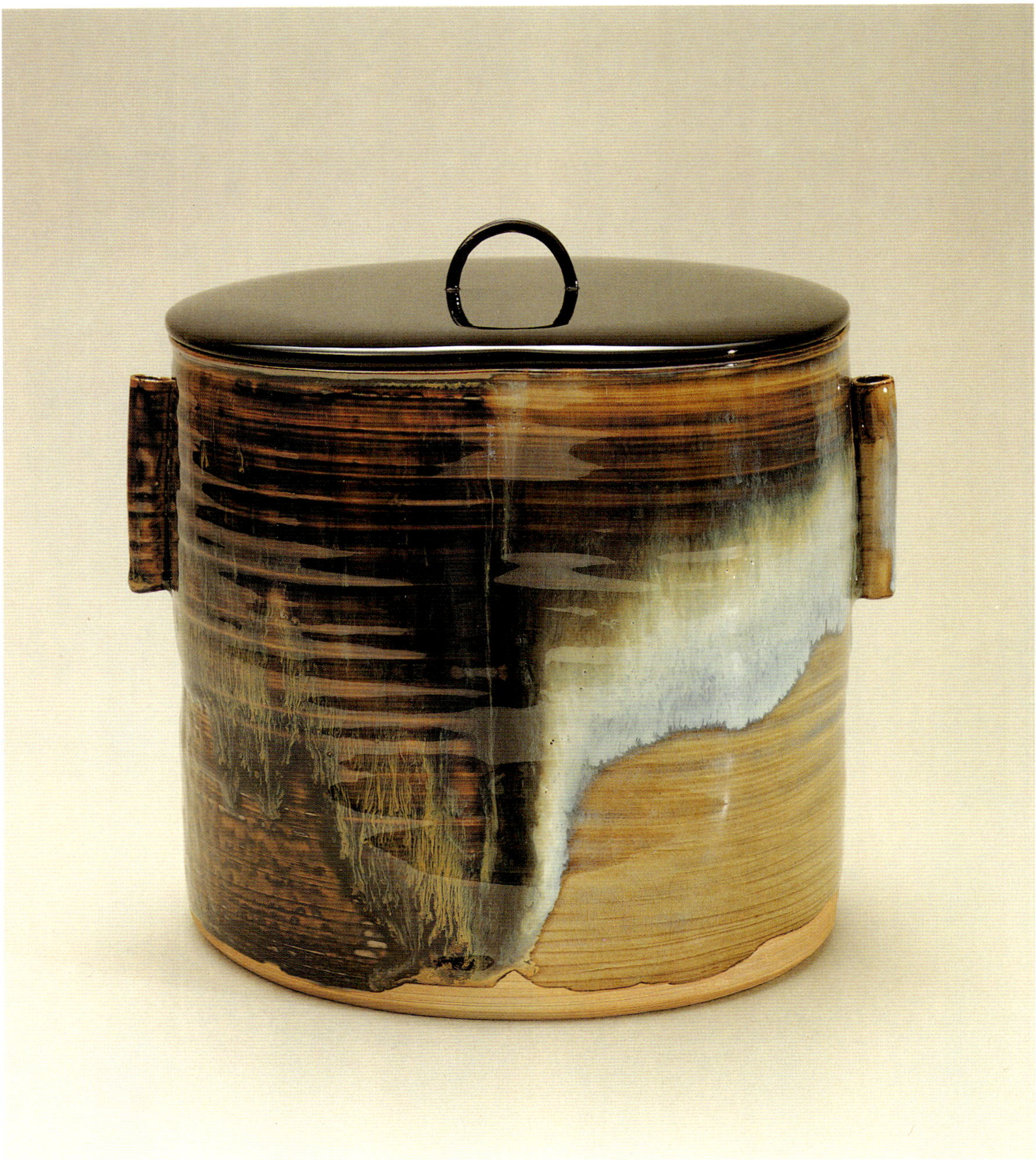

5. *Takatori ware: Water jar with multiple glaze pours, by Kamei Rakuzan (b. 1944). Ht. 15.0 cm. Private collection, Japan.*

Takatori ware was a vigorous country stoneware until around 1630, when its Korean founders traveled from their kiln center in present Fukuoka Prefecture to Kyoto to study with tea master Kobori Enshū. From that period, the products became increasingly refined. The chief technique in Takatori glazes is poured glaze decoration. Here a rice-straw ash glaze has been poured over an iron glaze. The horizontal incisions on the body provide an effective foil to the downward flow of the white accent.

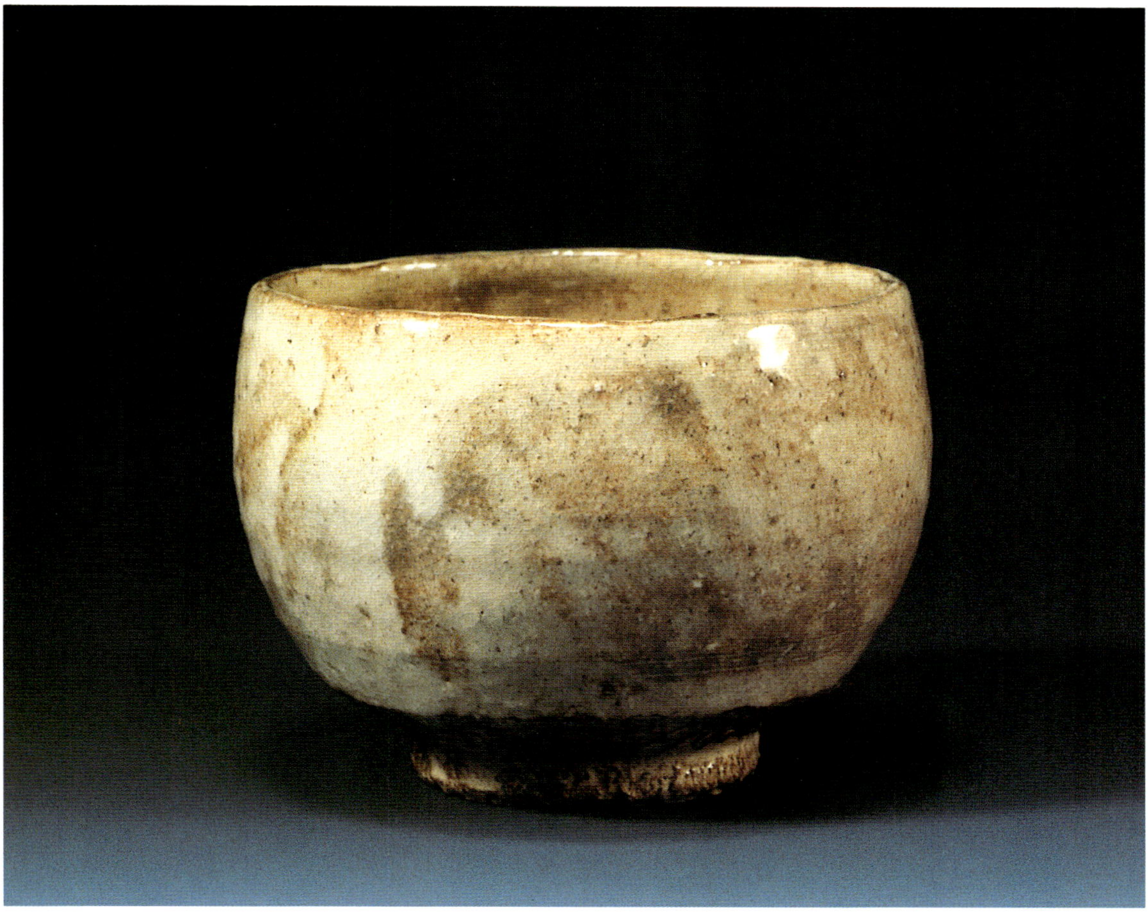

6. *Hagi ware: Tea bowl, by Tawara Tōbei (b. 1924). Stoneware, ht. 9.1 cm., diam. 12.2 cm. Private collection, U.S.A.*

The stoneware tradition that developed in Hagi, Yamaguchi Prefecture from the early seventeenth century is most famous for its tea ceremony bowls. The decoration relies on the simplest of effects—in this case, a casual and irregular application of white slip over a coarse reddish-brown body.

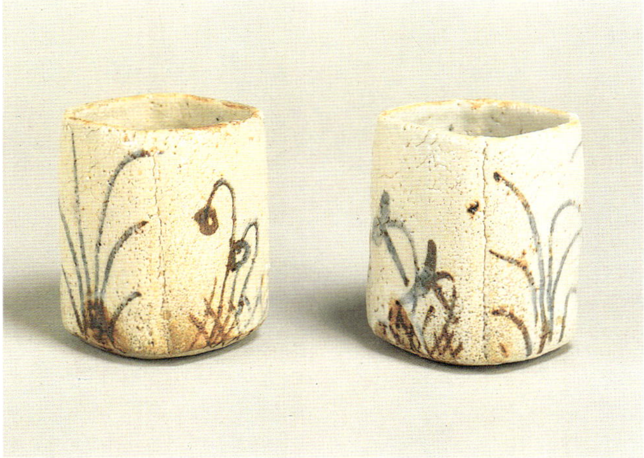

7. *Shino ware: Pair of food dishes with designs in underglaze iron. Late 16th c. Stoneware, ht. 8.3 cm. Private collection, Japan.*

Shino ware was produced in the Mino area of Gifu Prefecture. The vessels were thrown on the potter's wheel and then paddled into their square configuration. A vertical swipe from a rib then indented the corners. After being coated with a feldspar-ash glaze, the pieces were subject to a long firing in a damp kiln. The subtle pinkish body color, called "fire-color", results from iron in the clay migrating to the surface.

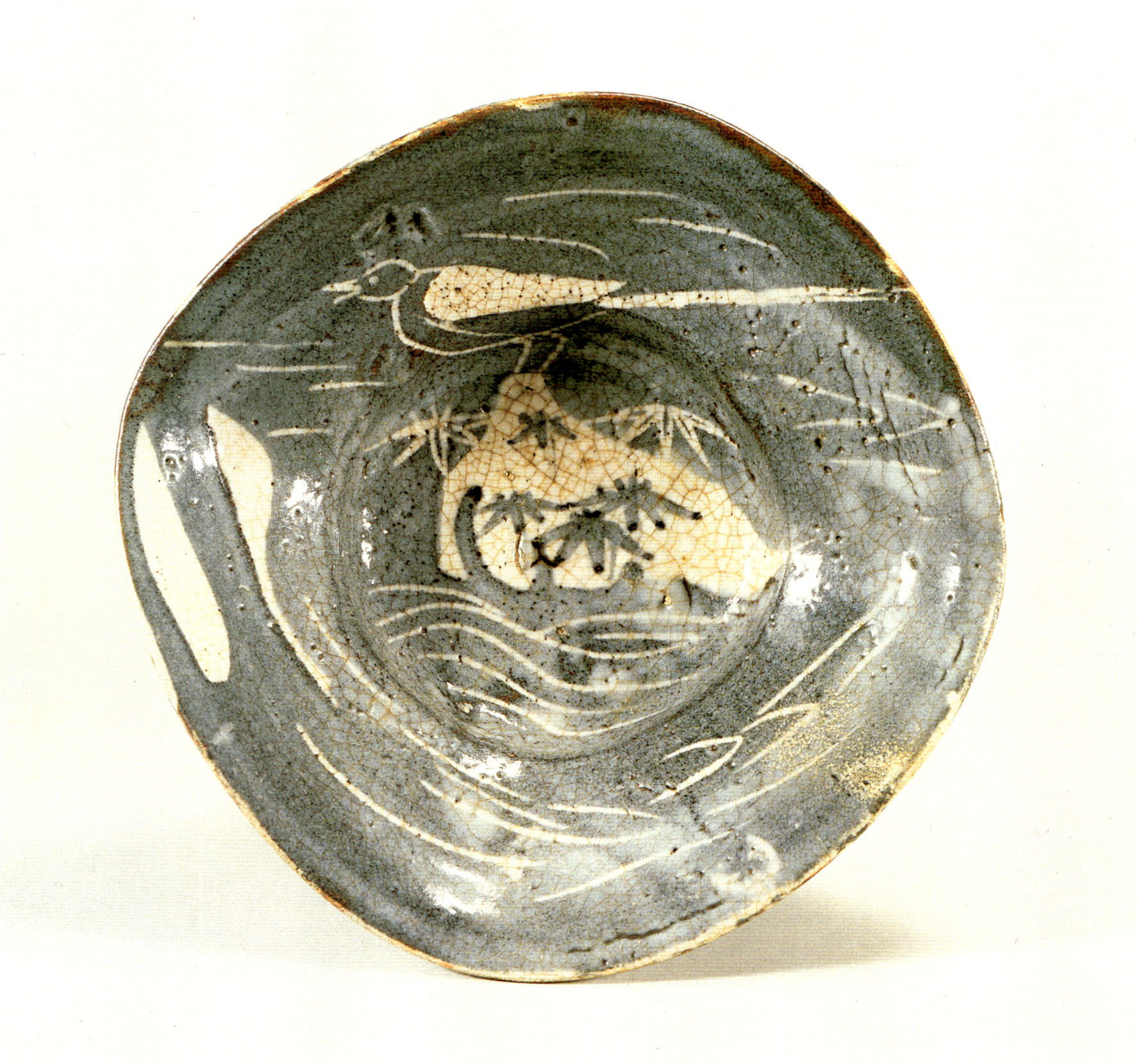

8. *Shino ware: Shallow bowl with design of wagtail on a rock in resist and sgraffito. Late 16th c. Stoneware, ht. 8.9 cm., diam. 26.8 cm. Tokyo National Museum.*

Shino potters applied a slip of their local iron pigment in the center, leaving randomly shaped white patches in resist. In this particular case, the resisted area suggested a rock, whereupon complementary details were engraved through the iron. This is the first use of sgraffito in Japanese ceramics history. The loose, spontaneous effect is complemented by the cloudy feldspathic glaze. A reduction atmosphere in the kiln has turned the body gray, and this variation is known as Gray Shino.

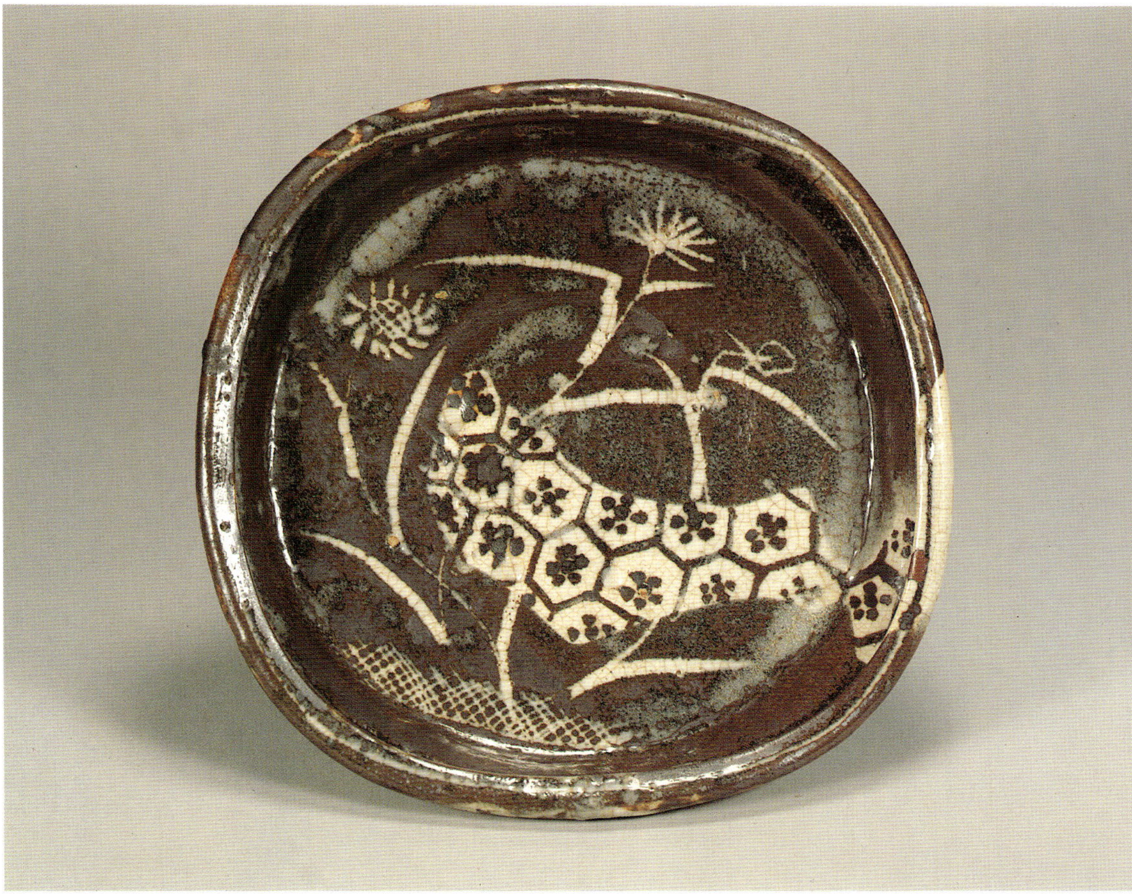

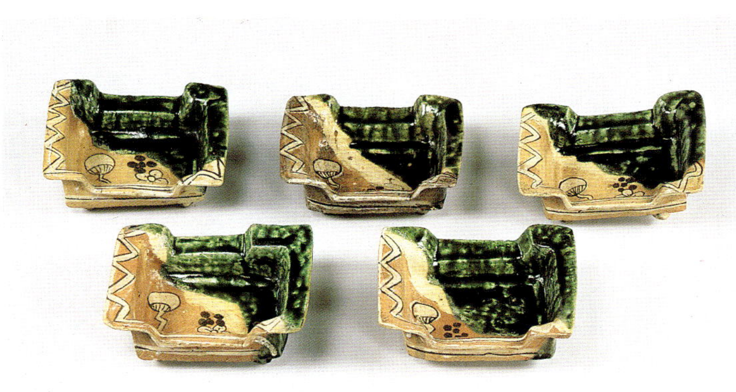

9. *Shino ware: Four-sided bowl with design of reeds and hexagons in sgraffito. Late 16th c. Stoneware, ht. 6.9 cm., diam. 26.8 cm. Gotoh Museum, Tokyo.*

The technique is identical to the bowl in plate 8, but this piece has been oxidized instead of reduced. The effect is known as Red Shino.

10. *Oribe ware: Set of food dishes with green glaze pours and underglaze decoration. Early 17th c. Stoneware, ht. 5.5 cm., w. 14.8 cm. Private collection, Japan.*

Oribe, named after a famous tea master thought to have influenced its style, was also made in the Mino district. The forms were drape molded and decorated with the local iron pigment. After coating the piece with a transparent glaze, the green glaze—the same transparent glaze with an addition of about 4 percent copper oxide—was applied as an accent. The piece was then fired in oxidation. This is the first use of a colored stoneware glaze (other than iron glaze) by Japanese potters.

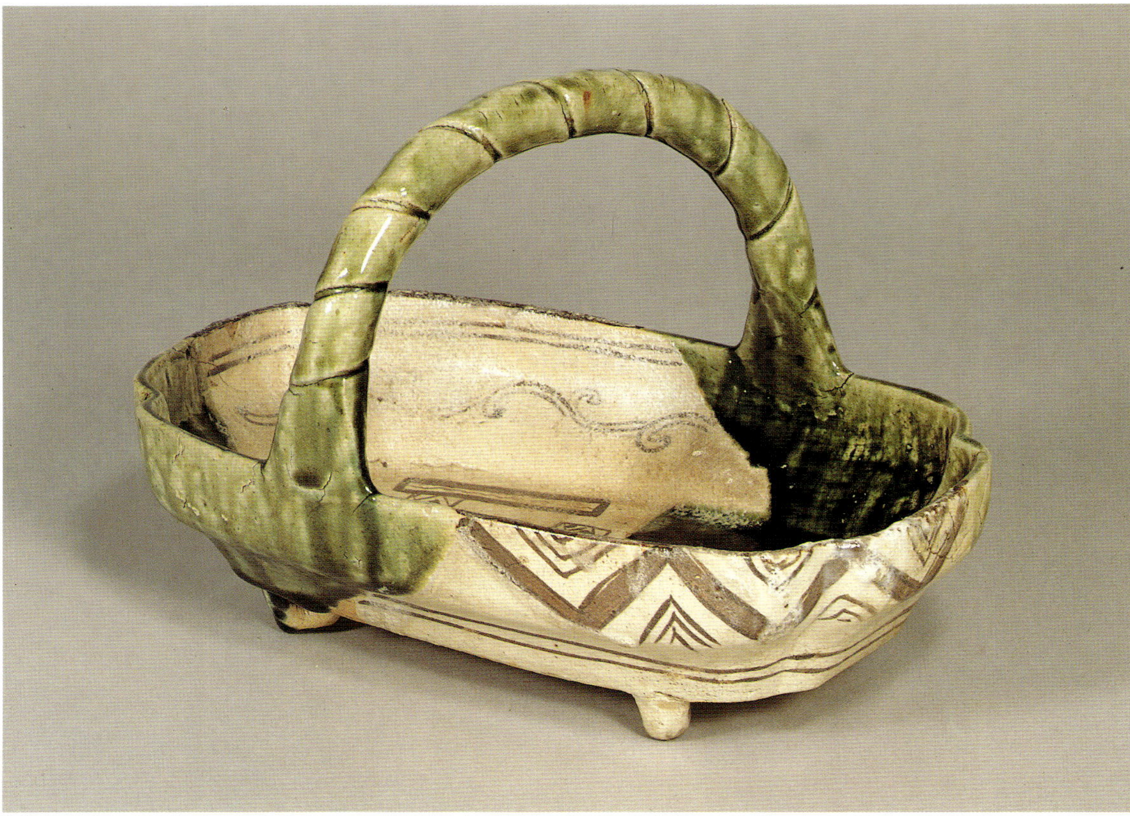

11. *Oribe ware: Handled dish with chevron and tendril design in underglaze iron. Stoneware, ht. 17.4 cm., w. 16.4 cm. Gotoh Museum, Tokyo*

The impure, slightly purplish pigment complements the pale green glaze accent on this piece. Larger Oribe ware food dishes are often equipped with a handle. These vessels were usually used to hold multiple portions of grilled food, or occasionally sweets.

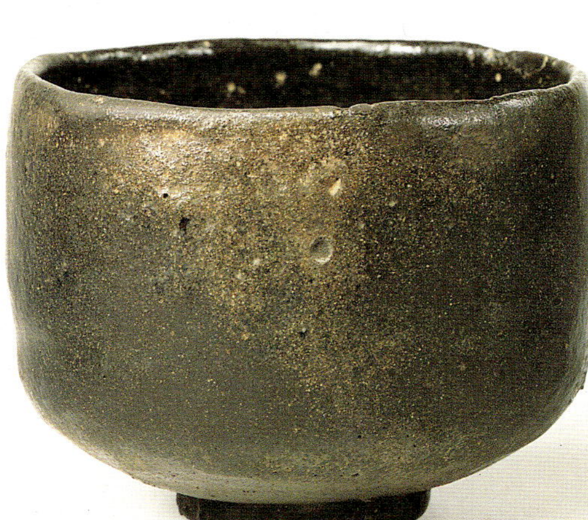

12. *Black Raku ware: Tea bowl, named Ōguro (Great Black), by Chōjirō (d. 1589). Late 16th c. Earthenware, ht. 8.5 cm., diam. 11.5 cm. Private collection, Japan.*

Raku ware is a product of Kyoto. This piece, typical of the style, is fashioned by hand and then trimmed. Its glaze consists of lead and Kamogawa stone, a ferrous rock. The bowl was extracted from the kiln at peak temperature, which turned the glaze black. The tong marks can still be seen. It is possibly the most famous Black Raku bowl in Japan.

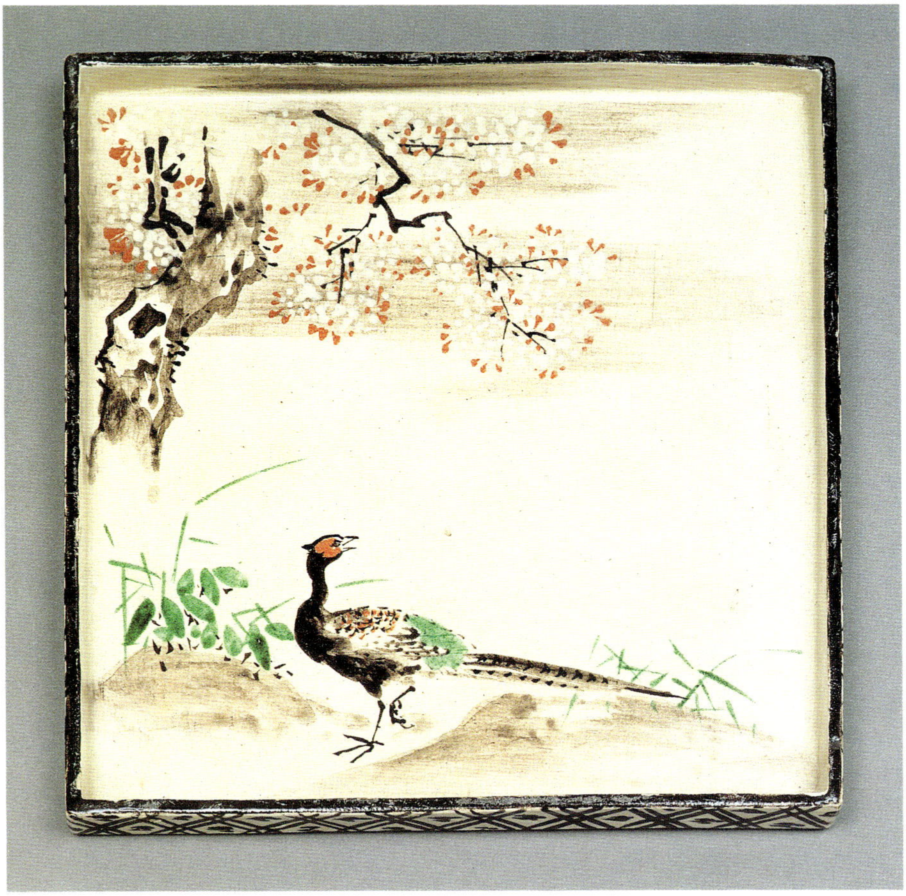

13. Kenzan ware: Square plate with bird-and-flower design in underglaze enamels, by Ogata Kenzan (1663–1743). Dated 1702. Earthenware, ht. 2.8 cm., w. 18.6 cm. MOA Museum of Art, Atami.

Flat geometrical shapes are typical of production in Kyoto, where Kenzan spent most of his career. These pieces were made by draping a thick slab over a mold. The slab was trimmed to its present thickness (a mere 4 mm.) over a prolonged drying period. Among Kenzan's many innovations in decoration was the use of pigmented lead glazes under a transparent base glaze. The colors blur in the firing, approximating the diffuse effect of painting on absorbent paper.

14. *Kenzan ware: Covered box with design of plum blossoms in underglaze pigments, by Ogata Kenzan (1663–1743). Stoneware, ht. 8.4 cm., w. 20.6 cm. MOA Museum of Art, Atami.*

The components were made over drape molds, and the sandy clay was scraped to produce a grainy surface. The decoration is inspired by stencil-dyed textiles. The white plum blossoms were first applied as a slip through a paper stencil, and then more plum blossoms were painted around them in iron and cobalt pigments. A reverse technique was employed on the inner surfaces, with the blossoms rather than the ground kept in resist. Kenzan delighted in adapting techniques and decorative effects from non-ceramic sources.

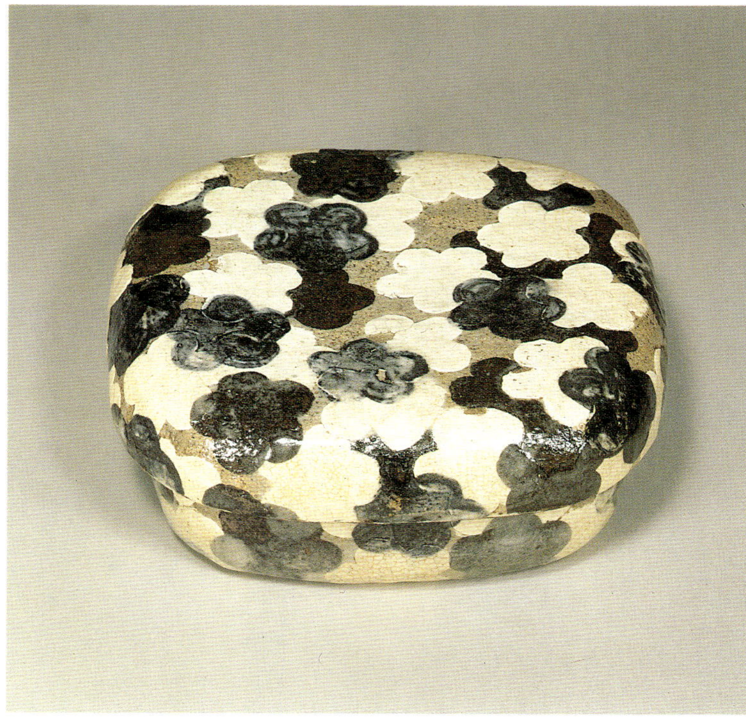

15. *Kyoto ware: Covered food dishes with striped designs in overglaze enamels, by Shirai Hanhichi (b. 1928). Stoneware, ht. 7.0 cm., diam. 13.4 cm. Private collection, U.S.A.*

An ordinary covered bowl shape has been stylishly modified by notching the rim. White slip is used to create a ground for the enamels, which are painted in linear patterns. These are intended to evoke the striped Western textiles that had an exotic appeal to premodern Japanese. The decorative effects and the rather soft stoneware firing with partial reoxidation around the glaze pinholes all contribute to the dreamy feeling typical of Kyoto stonewares.

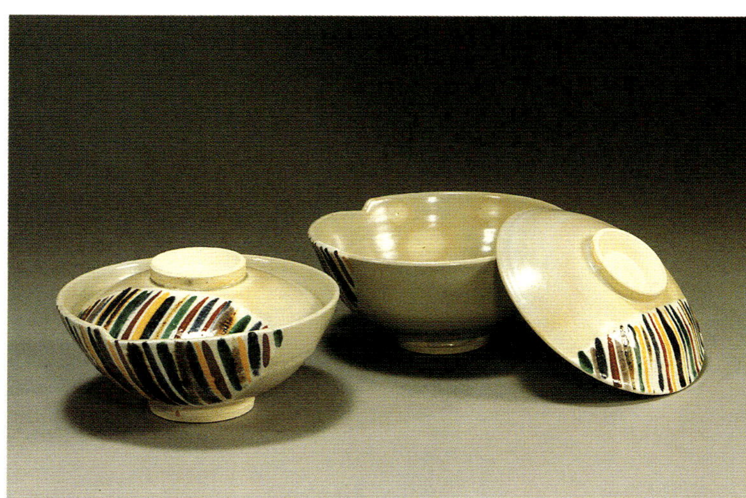

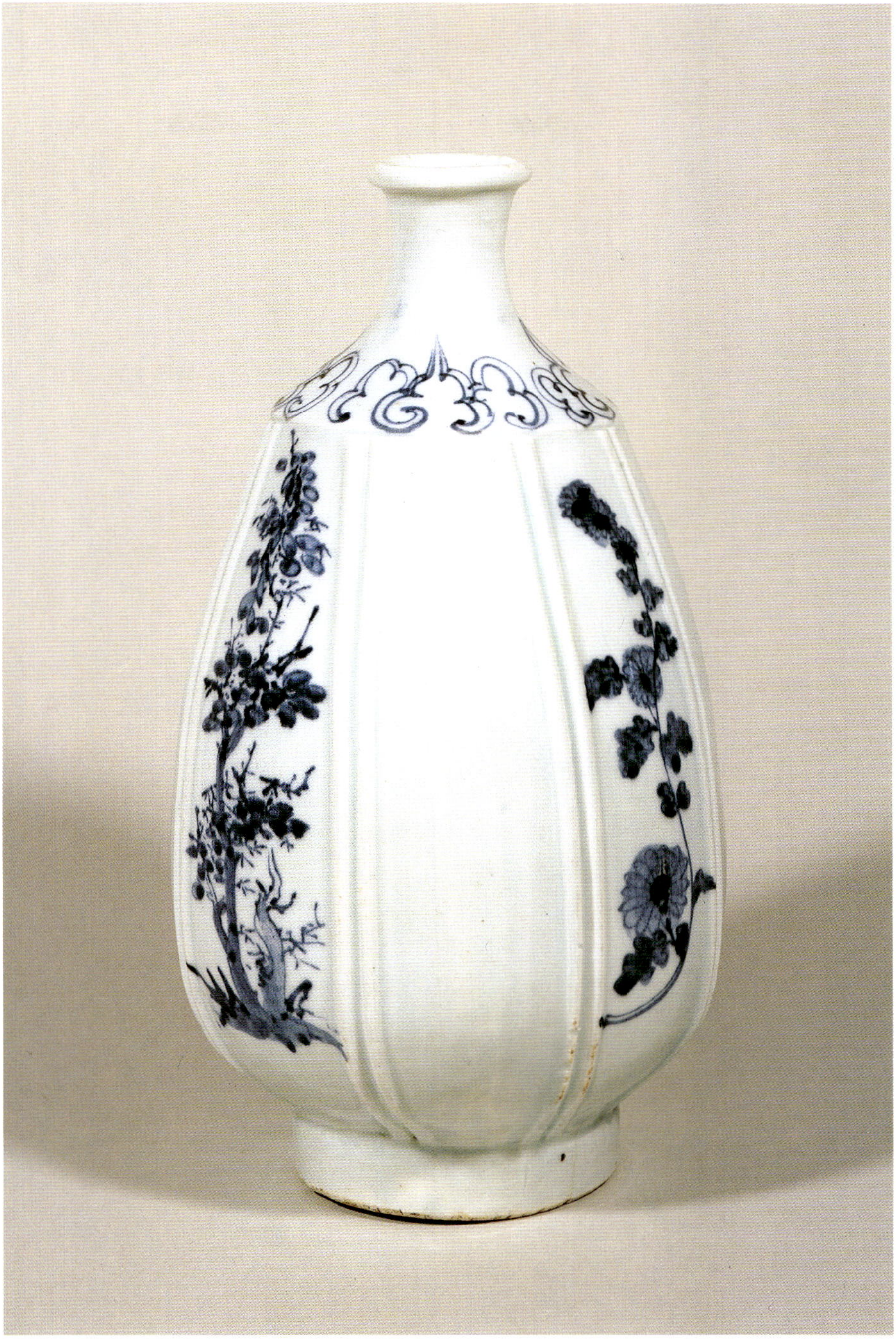

16. *Arita ware: Saké bottle with floral designs in underglaze cobalt. Early 17th c. Porcelain, ht. 32.0 cm., diam. 17.8 cm. MOA Museum of Art, Atami.*

This saké bottle was made not long after the Arita porcelain kilns were opened in the early decades of the seventeenth century in present-day Saga Prefecture. The walls of the bottle were thrown thickly, after which they were faceted and carved. Auspicious floral motifs painted in various shades of natural cobalt decorate alternate panels of the vessel. In early Arita porcelains like this one, all the decorating and glazing was done on greenware.

17. *Kutani ware: Plate with design of peonies in overglaze enamels. Mid-17th c. Porcelain, ht. 6.8 cm., diam. 37.0 cm. MOA Museum of Art, Atami.*

It was long thought that this colorful variety of porcelain was manufactured in Kutani village, present-day Ishikawa Prefecture, but recent archaeological and scientific studies have conclusively demonstrated that it was a product of the Arita kilns in Kyushu. Regardless of the origins, the rich enamel coloration and exuberant patterns have endeared Kutani ware to generations of potters and enthusiasts.

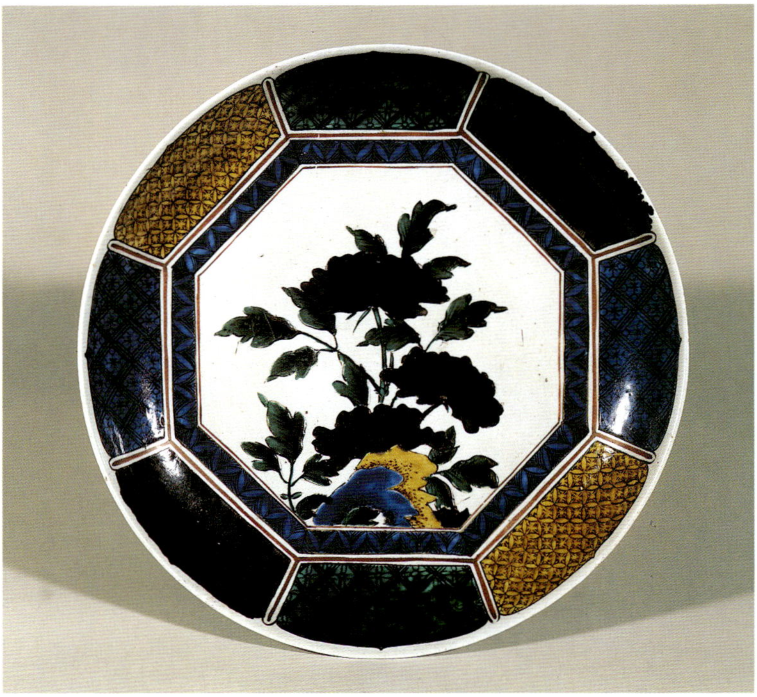

18. *Kakiemon ware: Lidded bowl with design of birds and flowers in overglaze enamels. Late 17th c. Porcelain, ht. 38.3 cm., diam. 31.6 cm. MOA Museum of Art, Atami.*

Since the early Japanese porcelain body was made of a pulverized stone with little plasticity, the forming and fitting of this large vessel required considerable skill. The decorative effect is dependent on fine enamel colors, used sparingly. The bright red is dependent upon a pure iron oxide (Kakiemon masters traditionally baked iron sulphate to obtain the high grade) and an excruciatingly long grinding. The Kakiemon workshop, located in the Arita kiln center, has been a leader in overglaze enameling since the mid-seventeenth century, and this is an early example of its style.

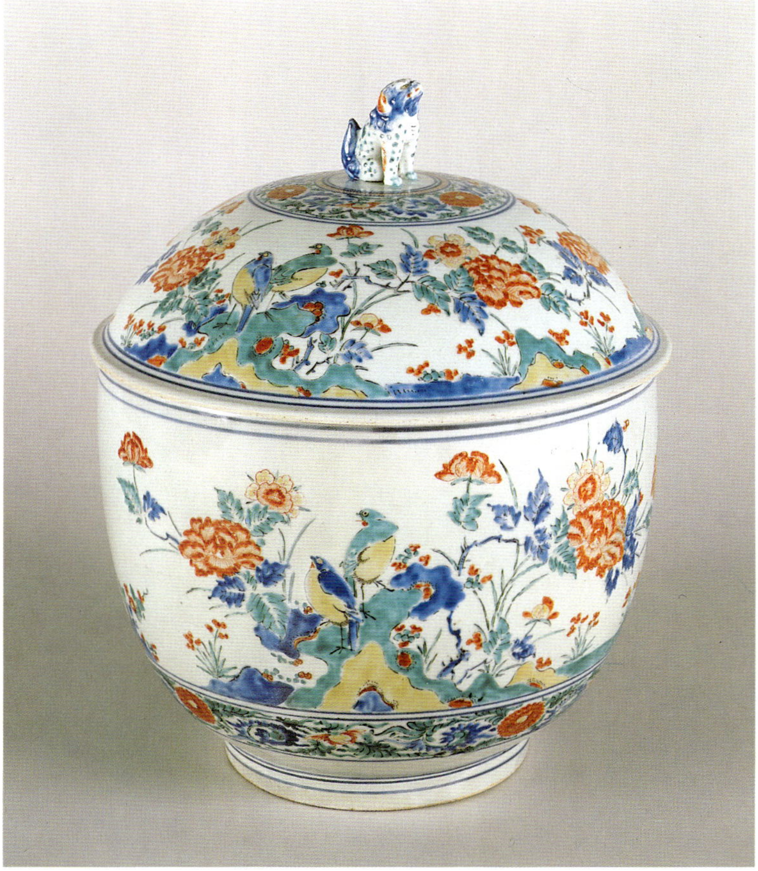

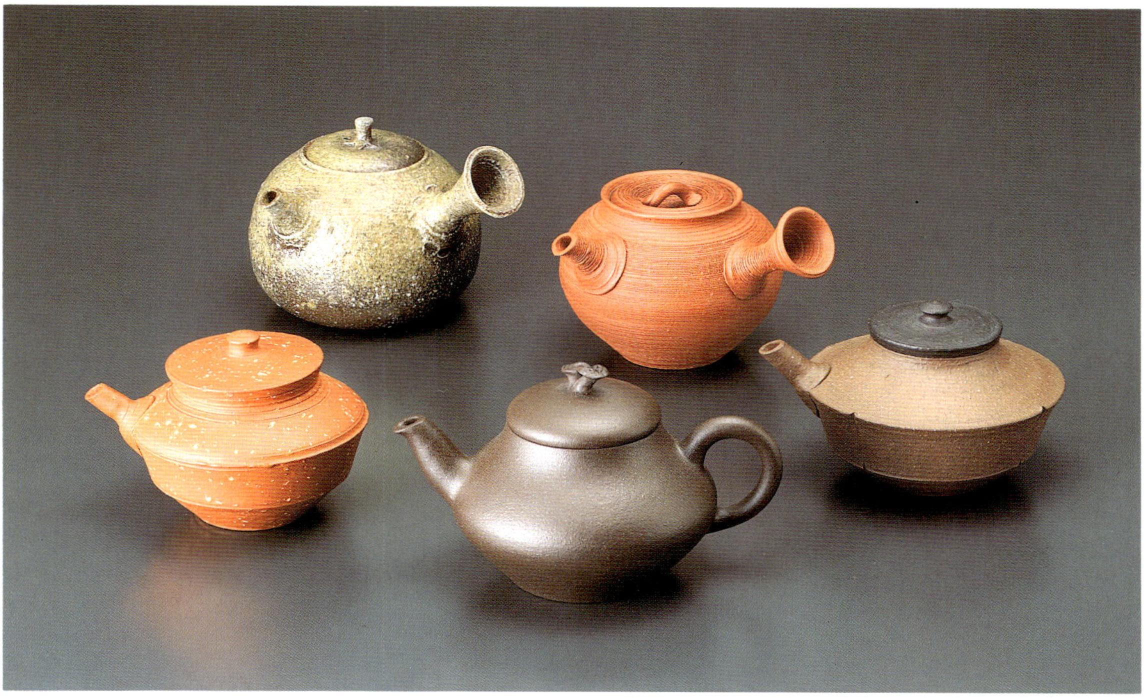

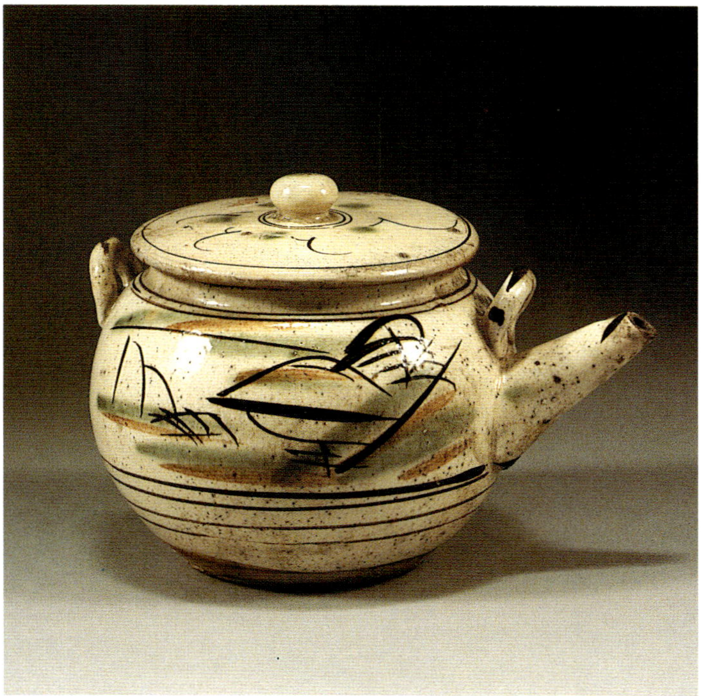

19. *Tokoname ware: Side-handled teapots (kyūsu), by Yamada Jōzan (b. 1924). Earthenware and stoneware, avg. ht. 6.0 cm., avg. diam. 10.2 cm. Craft Gallery, National Museum of Modern Art, Tokyo.*

Burnished teapots became popular in the nineteenth century with the growing fashion for steeped green tea. The fine red fusible clay of Tokoname (Aichi Prefecture) fired to a bright red in oxidation, and a brown in reduction. In addition to his mastery of the traditional smooth-surfaced *kyūsu*, Yamada Jōzan has experimented with textured clays, incised patterns, and natural ash glazing.

20. *Mashiko ware: Teapot with landscape design in underglaze iron and copper. Late 19th–early 20th c. Stoneware, ht. 12.0 cm., diam. 18.4 cm. Hachirō Yuasa Memorial Museum, International Christian University, Tokyo.*

This teapot is crisply formed out of a sandy local clay, covered with white slip, and painted in underglaze pigments. The transparent glaze was fired in oxidation to stoneware temperature. The skillful pattern painting is a triumph of repeat work. Mashiko, located in Tochigi Prefecture, arose in the mid-nineteenth century as a supplier of utilitarian stoneware for the nearby Tokyo metropolis.

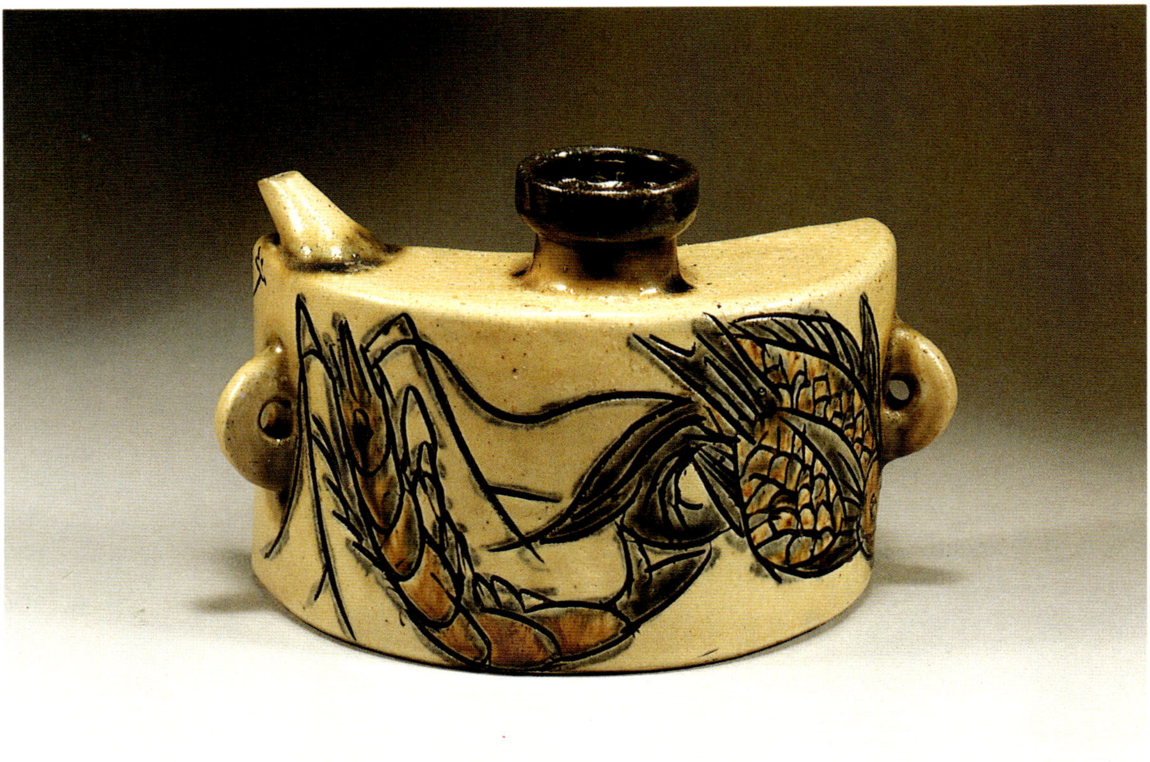

21. *Tsuboya ware: Hip flask (dachibin) with aquatic design in underglaze pigments, by Kinjō Jirō (b. 1912).Stoneware, ht. 16.3 cm., diam. 24.0 cm. Hachirō Yuasa Memorial Museum, International Christian University, Tokyo.*

Tsuboya ware, manufactured from the late seventeenth century, takes its name from a village now enveloped by Okinawa's largest city, Naha. The wares are characterized by distinctive shapes like this crescent moon-shaped flask, the subtle color of the off-white slip over an iron-rich body, and the fanciful painted and incised decoration. The tradition continues in Tsuboya and in the more rural village of Yomitan, where potters still fire with wood.

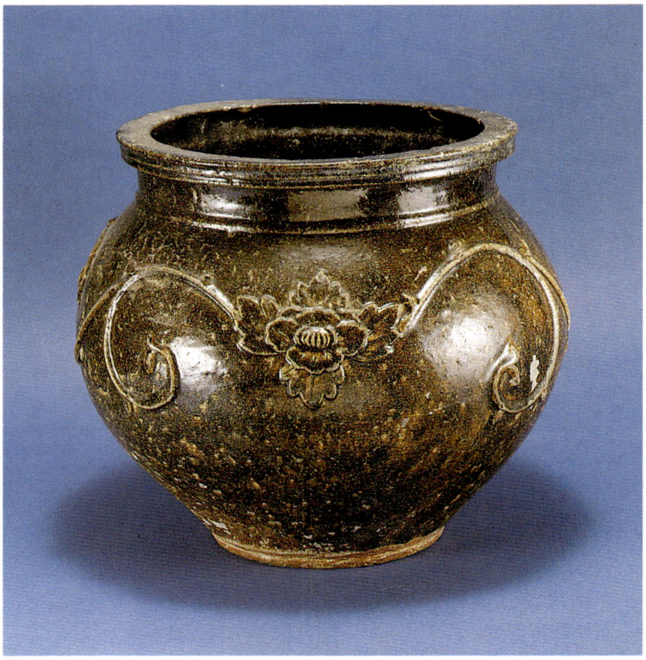

22. *Satsuma ware: Jar with appliqué designs of flowers and vines. Late 17th-early 18th c. Stoneware, ht. 31.5 cm., diam. 30.5 cm. Hachirō Yuasa Memorial Museum, International Christian University, Tokyo.*

The Satsuma potters (Kagoshima Prefecture), particularly those of Ryūmonji and Naeshirogawa, executed some of the most vigorous appliqué in the Japanese tradition. The black glaze is composed of a single material, a locally dug "glazing clay." The pot was used for fermenting rice in the making of saké.

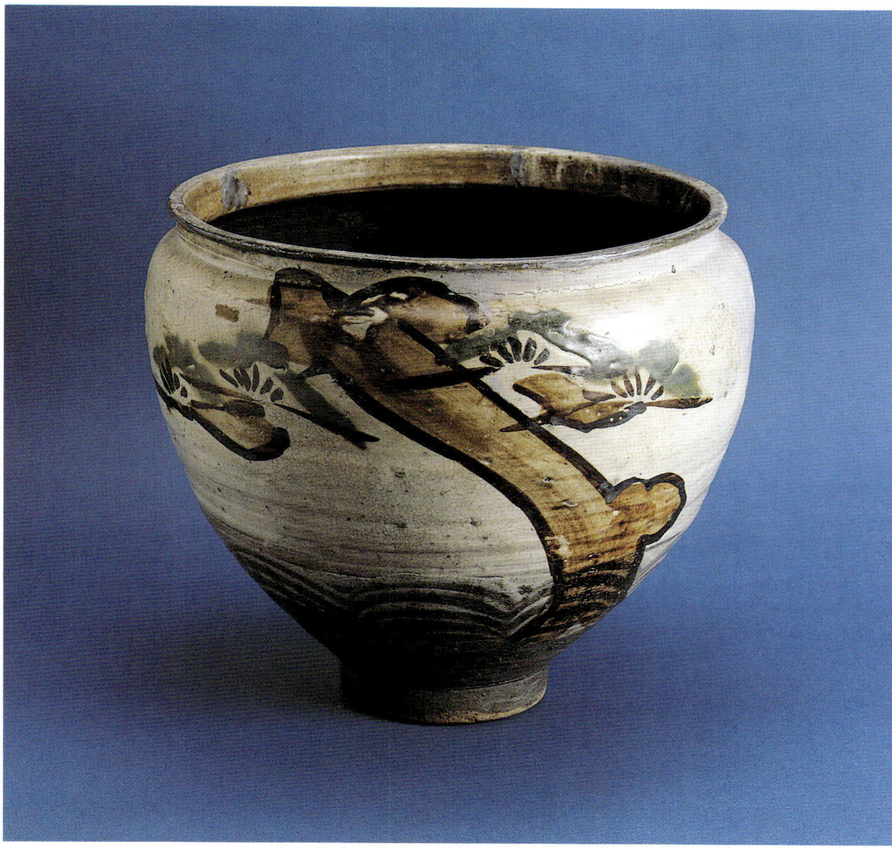

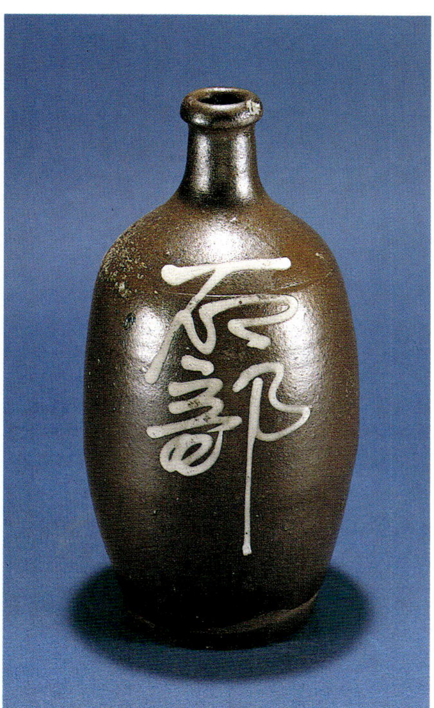

23. *Yumino ware: Jar with design of pines in underglaze iron and copper. Early 18th c. Stoneware, ht. 29.9 cm. Private collection, Japan.*

After being pushed out of tableware markets by the Arita porcelain kilns, many of the Kyushu kiln centers shifted to production for the farm and kitchen. The Yumino kiln, just south of the hot spring resort of Takeo, Saga Prefecture, was one such place. The off-white slip, finger-combed at the bottom to suggest waves behind the pine, combines wonderfully with the manganese-tinted iron pigment and the copper green. This type of ware is also known as two-color Karatsu.

24. *Tamba ware: Saké bottle with trailed slip design. Late 19th c. Stoneware, ht. 27.0 cm., diam 14.5 cm. Hachirō Yuasa Memorial Museum, International Christian University, Tokyo.*

The Tamba kilns, located in Hyōgo Prefecture, produced a wide variety of wares for everyday use, among them the vigorously slip-trailed bottles called "poor-man's saké flask." Each bottle bears the name of the wine shop, to which the empty bottles were returned.

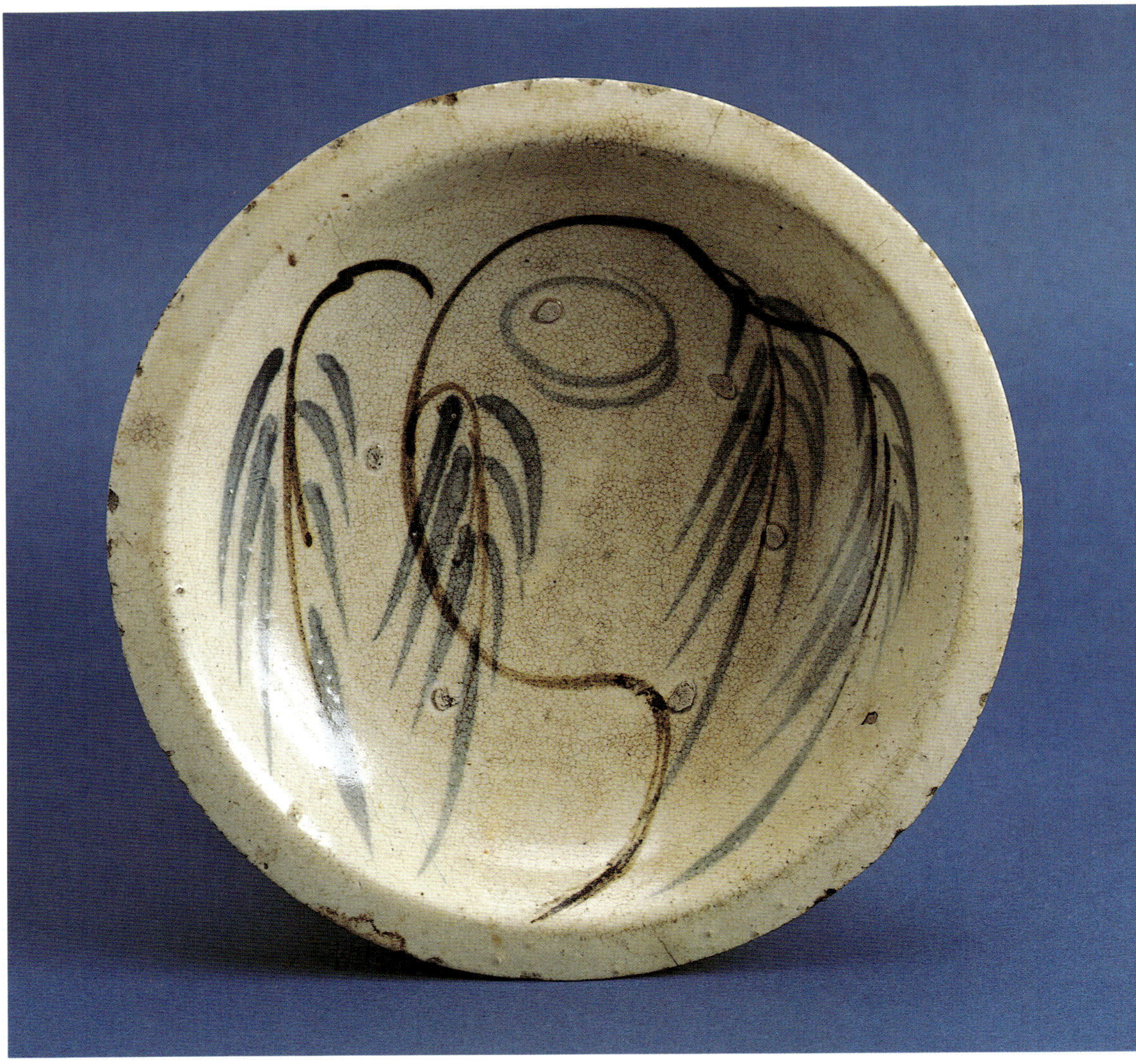

25. *Seto ware: Food dish with willow design in underglaze iron and cobalt. Late 18th c. Stoneware, ht. 5.2 cm., diam. 26.7 cm. Hachirō Yuasa Memorial Museum, International Christian University, Tokyo.*

These plates were produced in large numbers from the late eighteenth century at the Seto kilns, in present-day Aichi Prefecture. The natural cobalt and iron painting is renowned for its vigor and brevity. The bowls were used in food service for the common folk.

wares made by unselfconscious craftspeople possessed a selfless beauty akin to the Buddhist state of enlightenment. His movement also offered utopian alternatives to machine-age existence. At the grassroots level it renewed (or perhaps created) broad recognition for the beauty of the work of unknown craftspeople. In establishing museums and publications to showcase its aesthetics and by enfranchising retail outlets for contemporary folkcrafts, the movement surely saved numerous craft traditions from extinction. On the other hand, the most visible potter-exponents of the *mingei* movement such as Hamada Shōji (1894–1978; fig. 16) and Kawai Kanjirō (1890–1966) were not unlettered artisans but rather highly self-aware and urbane artists whose internationally known work fetched fine-arts prices. Tension between collective and personal causes, coupled with the inevitable social and economic changes that force formerly "unselfconscious" craftspeople to turn to media-driven mass markets, has come to plague the folk-art movement in recent years. A new generation of potters is either ignoring or rethinking Yanagi's exhortations to "health, beauty, and naturalness." Regardless of its contradictions, the *mingei* style became a major genre in twentieth-century Japanese ceramics and has contributed immeasurably to the diversity of contemporary Japanese crafts. Through the writings of a non-Japanese participant, British potter Bernard Leach (1887–1979), *mingei* has for many Western audiences become synonymous with Japanese ceramics as a whole.

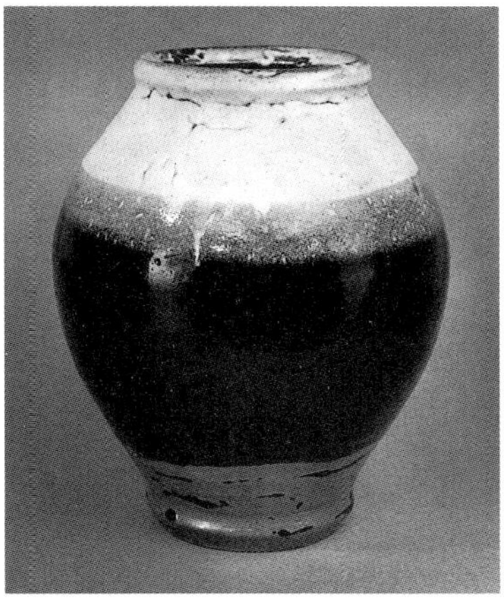

16. Jar, by Hamada Shōji

One target for subtle derision by the *mingei* movement was an older and more conservative tradition of family potteries, which had survived though fixed patronage or specialization in a certain product, particularly wares for the tea ceremony. But the tea-ware specialists had also suffered under modernization. A typical example is Nakazato Muan (1895–1985), a twelfth-generation scion of a family of Karatsu-ware potters, which, until the collapse of the feudal system in the mid-nineteenth century, had enjoyed the support of the local military government. By the time Muan (then called Shigeo) became head of the family in 1927, the workshop had failed after a period of marginal survival in making ornamental sculpture. But an opportunity to excavate older kilns in the area two years later stimulated Muan to experiment with reproducing classic Karatsu tea-ware

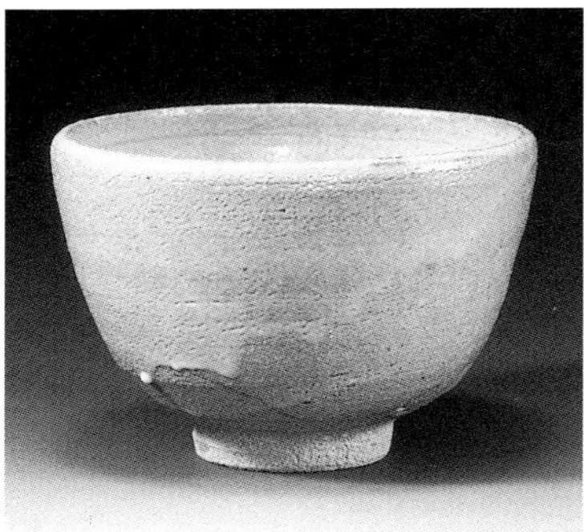

17. Tea bowl, by Nakazato Muan

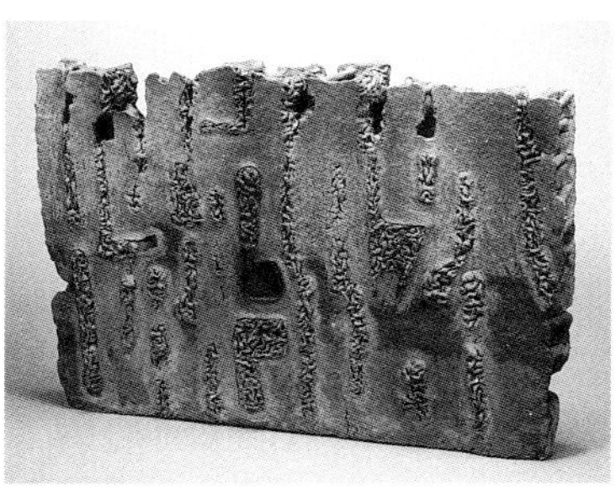

18. *Hekitai* (Wall Body), by Yagi Kazuo

styles from the heydays of the late sixteenth and early seventeenth centuries (fig. 17). A resurgence in the popularity of the tea ceremony further stimulated those efforts. Elsewhere in Japan, Arakawa Toyozō (1894–1985) revived Shino ware, Miwa Kyūwa (1895–1981) revived Hagi ware, and Kaneshige Tōyo (1896–1967) did the same for Bizen ware. In the postwar period, when interest in all traditional ceramics boomed, a number of potters in the *mingei* and tea-revival styles were honored as holders of intangible cultural assets (the so-called "national treasures"). In addition to the *mingei* and tea-ware practitioners, there are numerous other potters working in more or less traditional styles. Some of them have become independent after a period of apprenticeship in an established pottery, and others, especially graduates of art schools, have joined one of the numerous craft societies. Many Japanese potters tend to make career-long affiliations of this nature, and more often than not their work is known and appraised in that context.

This volume is concerned primarily with the techniques of traditional Japanese ceramics, but it would be remiss not to mention the thriving contemporary scene, much of which is dedicated to work of a nonfunctional nature. Sōdeisha, a group founded in 1948 by Kyoto potters Yagi Kazuo (1918–1979; fig. 18), Yamada Hikaru (b. 1924), and Suzuki Osamu (b. 1926) continues to be a major influence. An exploration of such contemporary masters, many of whom work in Western-derived methods, will have to await a later study. Yet even though their styles represent a departure from tradition, an insistence on simple forms, exacting craftsmanship, and an exposition of primary materials serves to link them with centuries past. In Japan, as in all of East Asia, tradition carries a positive connotation as a source of artistic guidance and spiritual regeneration.

NOTES

1. Based on *Tsuchi to Honoho* [Clay and fire] (Saga: Kyushu Ceramics Museum, 1981), 1.

2. The author recently witnessed the same thing when a folk art authority gave a lecture on Indonesian ikat textiles at a Tokyo museum. The room hummed with interest as various samples were passed around, to be lovingly seen, touched, and even smelled!

3. Kuroda Ryōji, *Shino,* trans. Robert N. Huey (Tokyo: Kodansha International, 1984), 14.

TWO

Setting Up

Good lessons, according to an old Japanese saying, proceed from one to ten—and then back to one. That is, the attainment of skill or knowledge always involves returning to the basics. It may be simplistic to assert that great pots are born from fundamentals, but it is indisputable that through attention to the basics, craft becomes sharp, rooted, and sure. This belief is in turn subtly imparted to users: sure manufacture makes for sure use.

Cleaning the workshop, repairing the tools, and making a fresh batch of clay may seem like menial apprentice chores, but they also provide times of repose for the master potter.

WORKSHOP

Workshop setting, minus large kiln, raw materials storage, and damp room (fig. 19). Potters often have little control of the structures that they occupy, but a good workshop layout, one that takes into consideration the relationships among various procedures, saves much unnecessary labor. The basic fixtures of the Japanese workshop are very convenient and congenial to small-scale production. Windows usually face south, providing ample light (usually supplemented by a bare incandescent light overhead) for working at the potter's wheel. Ware racks are out of the direct sunlight. The space shown here is 8.3 meters on the long side and 6 meters on the short.

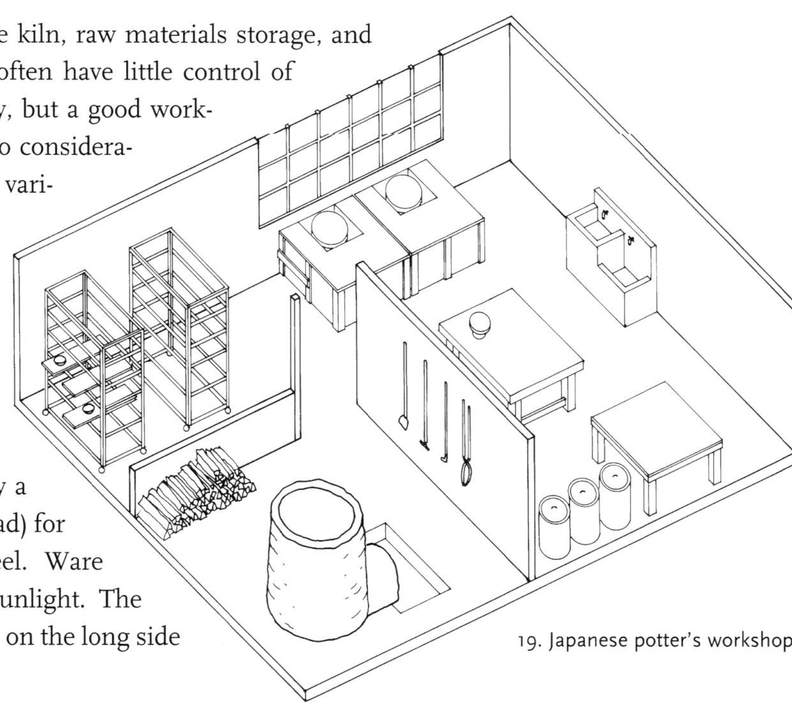

19. Japanese potter's workshop

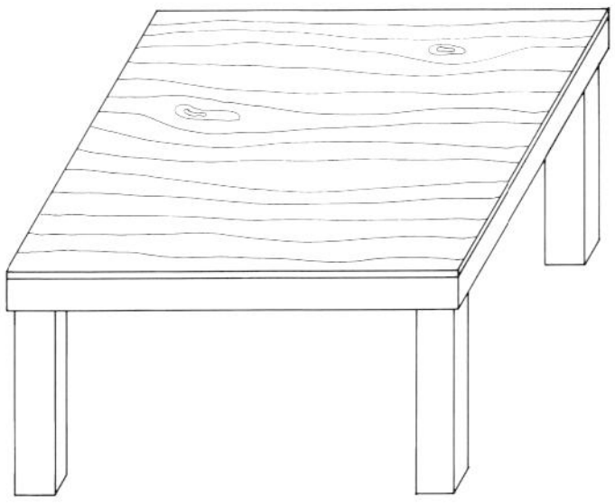

20. Kneading table

The kneading table (fig. 20) is sturdily built. Its surface is slightly absorbent; most potters use plywood or plank tops. Height is calibrated so that the tabletop just grazes the fingertips when the potter stands hands at sides. This may seem low, but this height brings more body weight into the kneading. The tabletop is at least 80 x 80 centimeters in surface area.

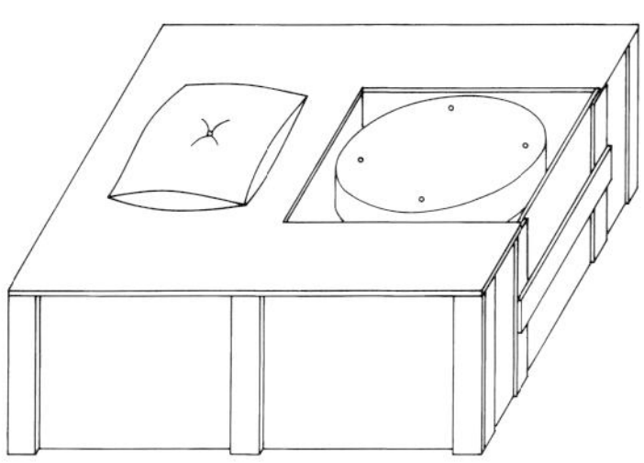

21. Potter's wheelbox

Setup for the potter's wheel (fig. 21). This arrangement situates everything—tools, water, ware boards—within easy reach. The box-like enclosure has the advantage of catching the shavings that spin off the wheel during trimming. They are cleaned out periodically by opening up a removable side of the box. Hand-wheel potters and potters using an electric wheel with a stick lever may sit cross-legged on top of the platform. Otherwise, the area in front of the wheel is opened up, and the potter sits on a stool or bench. The fixture shown here measures 120 x 120 centimeters in working surface, and is 40 centimeters high. The wheel well is 60 centimeters square. For more on the potter's wheel itself, see page 72.

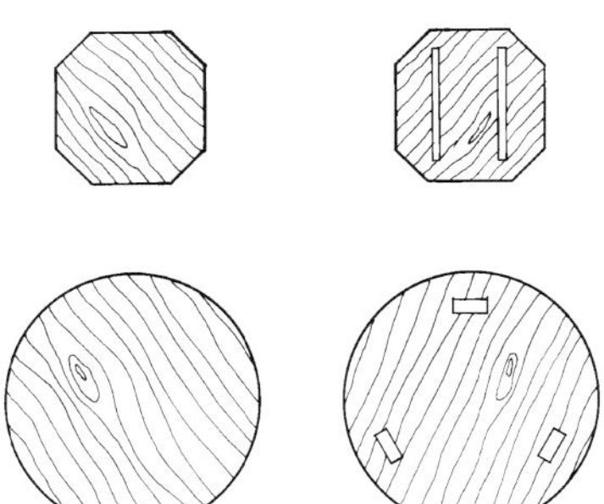

22. Bats

Bats (fig. 22) make it easy to transfer large pots off the wheel. They are also necessary for making sections of pots for later assembly and for trimming pots whose diameter is larger than the wheelhead. Eight-sided cleated bats, usually about 24 centimeters wide, are the most common. They are fixed to the wheelhead by coils of clay (see page 80). But as an alternative to these traditional bats, which are hard to center and recenter, a plywood disk cut to a diameter larger than that of the wheelhead, about 45 centimeters, can be used. Three short cleats are attached to the underside of this bat. They are positioned so that when the bat is placed on the wheelhead, the cleats will abut its side. The bat is tightly secured to the wheelhead by wedging a sliver of wood (disposable chopsticks work well) between one of the cleats and the side of the wheelhead. When the bat is removed with work in progress, note is taken of which cleat was wedged, and that same cleat is wedged when the work is returned to the wheel. It is very easy to recenter work with this bat.

Ware racks (fig. 23) are widely used in workshops and in art schools. In older workshops the vertical supports are wood and the horizontals are sturdy bamboo; the rack is either attached to the wall or suspended from the ceiling. An angle-iron rack on coasters is a frequent contemporary replacement. A standard size is 170 centimeters high, 140 centimeters long, and 65 centimeters deep. There is usually 30 to 40 centimeters of space between each tier of shelves.

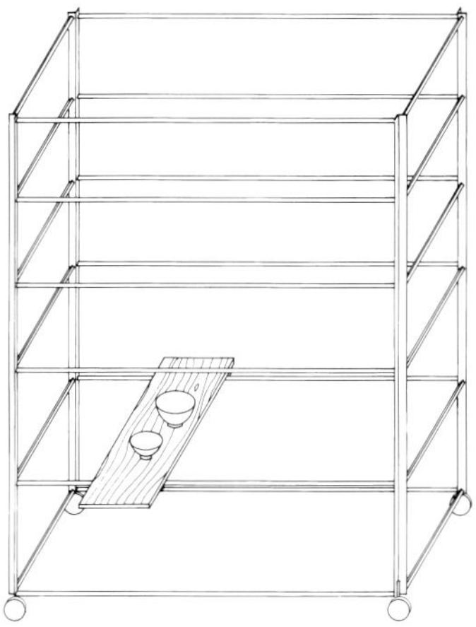

23. Ware rack

Ware boards (fig. 24) are cheap softwood planks, quartersawn to prevent warping. Plywood is practical where the humidity is low. The boards are cleated for easy pickup. A standard size is 100 centimeters long by 25 centimeters wide. The length of the boards and the placement of their cleats are, of course, adjusted to the ware rack.

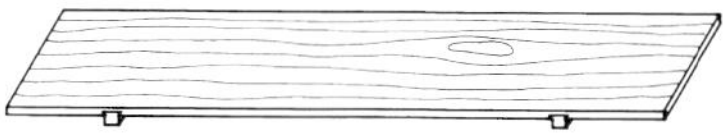

24. Ware board

A good sink (fig. 25) is a back saver. Many workshops contain two sinks, a high one for washing small items like tools and a low one for large items like buckets and bats. Inexpensive sinks are made from poured concrete coated with waterproof paint. Drain plumbing is accessible. The one pictured here is 90 centimeters high (to top of backboard) and 120 centimeters wide. Both sinks project 45 centimeters from the wall and are 20 centimeters deep. The sinks are 50 centimeters and 75 centimeters high, respectively, and the walls are 8 centimeters thick.

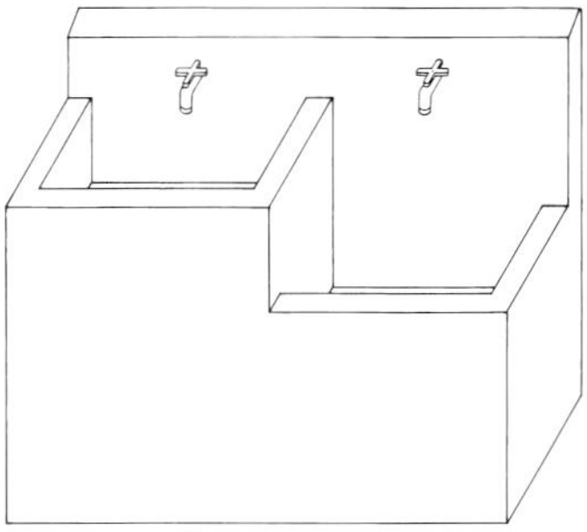

25. Sink

TOOLS

Pottery apprentices and even art school students in Japan learn how to make their own tools. This is not merely an exercise in humility. To begin with, specific tasks call for specific tools. As a potter's product line evolves, tools such as ribs and brushes are tailored to suit particular designs. Tools also vary as a function of local custom. Potters in Kyushu tend to use a long, tongue-shaped forming rib, and those in Kyoto use a shorter, rounder one. Raku potters use a set of about ten trimming knives for specific stages in trimming their tea bowls; other potters use only one or two. Finally, tools are a matter of personal preference and physical convenience. The tendency of Japanese potters to design and redesign their tools shows

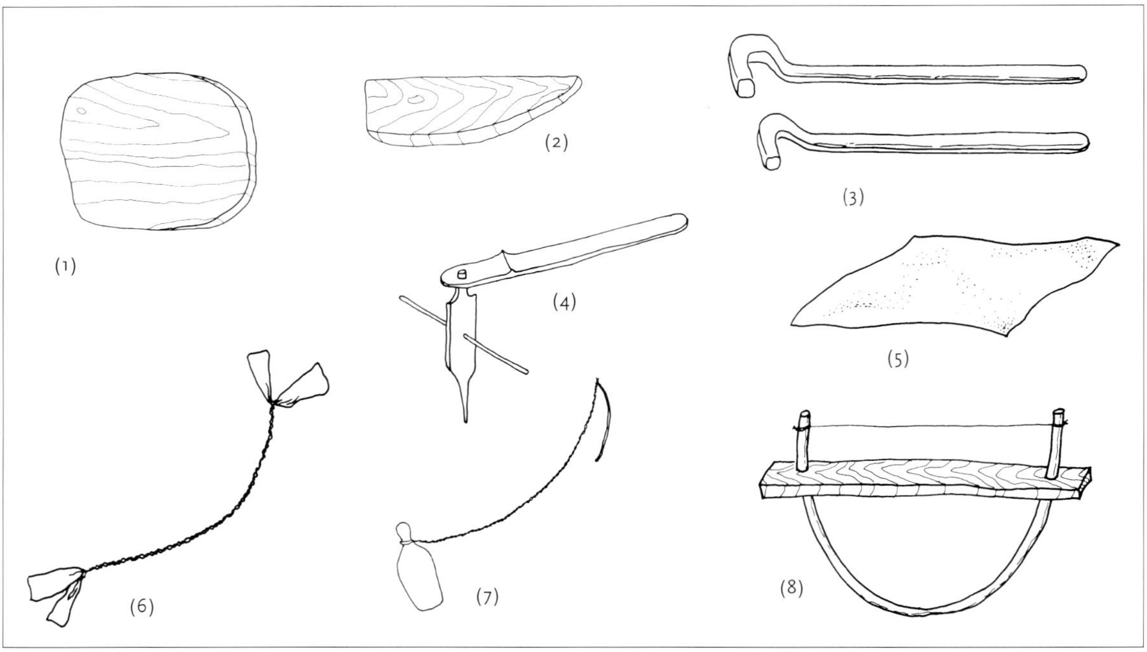

26. Forming tools for use on the wheel

how good craftsmanship is found at every stage of the work. As the potter becomes intimate with tools, he or she also becomes familiar with their creative potential. Naturally, some of the tools used in the West and in Japan are similar, but here, in four categories, are some distinctive Japanese tools.

Forming tools for use on the wheel (fig. 26). The forming rib (1) is thick and blunt, and usually made of softwood. The finishing rib (2) is thinner and sharper, and usually made of hardwood. The throwing stick (3) is used in throwing and shaping hollow ware, especially bottles. The "dragonfly" measuring gauge (*tombo*; 4) is a bamboo tool that the potter uses to measure simultaneously the width and depth of a pot. One is usually made for each model in the production line. The handle of the gauge is sometimes sharpened to a fine point, in which case it also serves as a trimming knife. A chamois (5) is used to smooth edges of pots. The cutting string (6 and 7) is used to cut the pot off the bat or a larger clay mass. These are usually braided cords or twisted wires. The trimming bow (8) maintains a taut line for cutting edges.

Trimming tools (fig. 27). Bottomless wheel-thrown cylinders, or chucks, hold the pot while it is being trimmed. Pictured here are chucks for cups and small bowls (1), hollow ware (2), and flatware

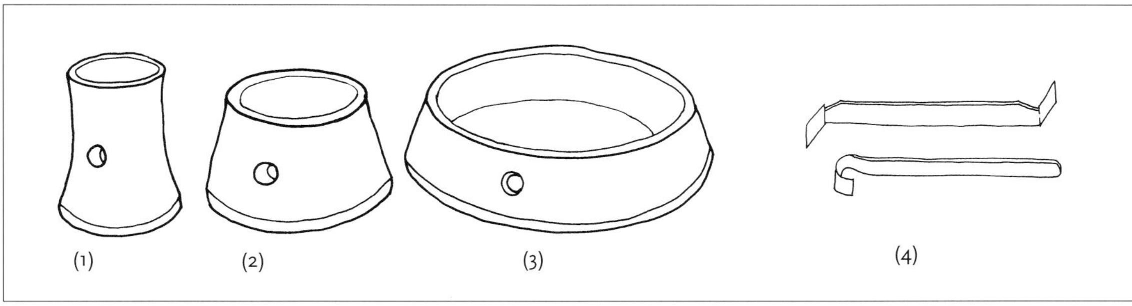

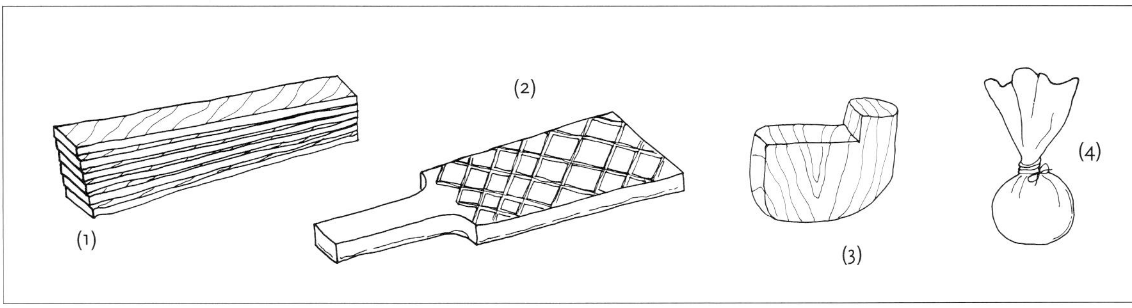

27. Trimming tools

28. Forming tools for coil and mold-made pots

(3). See pages 93–94 for more about chucks. Trimming knives (4) are typically used for trimming excess clay from the bottom of thrown pots. They are made of banding iron, cut to size and then bent in a vise. Blades are shaped with foot ring and base contours in mind. The cutting edge is on the outward-facing side of the tool and is kept sharp and free of rust. Traditionally, potters heated the blades before sharpening them.

Forming tools for coil- and mold-made pots (fig. 28). Measuring sticks (1) are long, flat sticks (usually 45 x 3 cms) that serve as guides for cutting slabs from a lump of clay. They are also useful as straight edges for cutting out slabs and as reinforcements while luting walls. A clear-grained wood, one resistant to warping, is preferred. Thicknesses of 5 and 7 millimeters are the most common; this measure will determine the thickness of the cut slabs. Paddles (2) are used for compressing forms and sealing joints. When the paddles are textured, they also add an element of surface decoration. An anvil (3) supports the inner wall during paddling. Dust bags (4), usually made of cotton cloth, are used in press and drape molding. One type, filled with powdered starch, is used for dusting molds prior to use; the other, filled with sand or grog, is used for patting slabs into place while molding. The size depends on the job. The bags are closed tightly when in use.

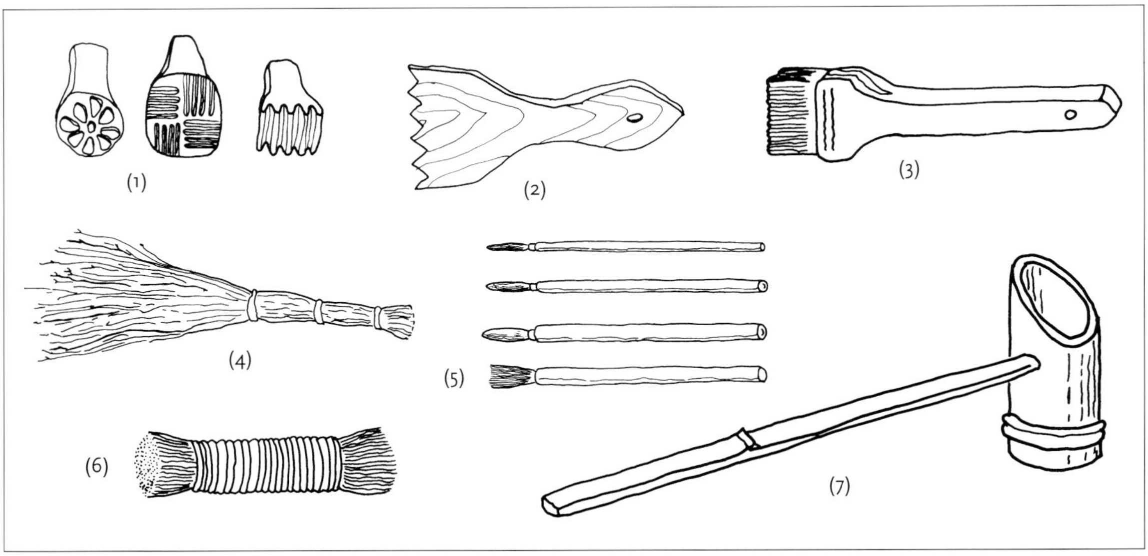

29. Decorating and glazing tools

Decorating and glazing tools (fig. 29). Bisque-fired stamps (1) are used for impressed designs; such designs can be repeated in a band using patterned rollers, or roulettes. Combs (2) are used for incised decoration in parallel bands or waves, and are also convenient for cross-hatching surfaces prior to luting. Flat "chisel" brushes (3) are used for applying slip and glaze. They are usually made from animal hair. Rough straw or twig brushes (4) are used for coarser applications, especially the Korean-inspired slip application called *hakeme*. Such brushes are no more than a bundle of fibers, and effects vary according to their rigidity. Pointed brushes (5) come in many varieties. Potter's brushes are similar to those used in painting or calligraphy, except that they are made of stronger, coarser hairs (deer, for example) that can hold the relatively dense pigments and glazes used in ceramics. For washes between bounding lines, especially in cobalt pigment, a fat brush with a reservoir in the handle is used. Pigment is released by squeezing the brush tip. Scaling brushes (6), made from bundles of rigid palm fibers, are very convenient for cleaning glaze from bases and foot rings prior to firing. Ladles (7) are used to apply glaze and slip, the size depending on the job. The particular advantage of the Japanese ladle is its angled mouth, which makes pouring easy to control.

This is a very brief description of tools used by the Japanese potter, and it is intended only as a guideline. There are two key principles: tools should match the job; and tools should have their own particular expressions. Fancy, store-bought tools may be easier to operate at first, but these traditional tools are easy to make, maintain, change, and, in the long run, they are easy to use.

CLAYS

Writing in 1897, Japanese art enthusiast Justus Brinckmann criticized the Western connoisseur's obsession with the shapes and surface decorations of ceramics and decried their neglect of "the vessel's body, the material from which it was made, and the technical manipulations which ensue from its forming."[1] Brinckmann had struck on the central truth that clay is the first point of attraction for the lover of Japanese pots. It is also a prime focus for the Japanese potter. Indeed, the potter's extraordinary attention to basic materials is the single largest difference between Japanese and non-Japanese techniques.

While looking for clays and blending them, potters typically search for qualities that enhance forming (high plasticity, low shrinkage) and firing (good fired strength). The potters in Japan are no exception. Pottery clays found in the Japanese archipelago range from low-firing earthenwares to refractory kaolins. From the Neolithic (the Japanese Jōmon) period through the early centuries of the common era, the potter relied on simple iron-bearing shale and alluvial clays of a type that is generally classified as earthenware. Preparation was confined to addition of temper. Organic materials were used as temper in much early Neolithic work, but sand or crushed stone predominated thereafter. Further refinements developed under Chinese influence in the eighth and ninth centuries, when potters of Nara three-color ware and Heian ash-glazed wares sought out white, refractory clays and enhanced their fineness through washing, or levigation. Little effort was made to refine the stoneware clays used by the numerous rural kilns that arose in the medieval period. In the seventeenth century, however, with the discovery of porcelain stone in Japan, smooth, fine clays became the general standard.

In contrast to this trend towards refinement, the peculiarities of many regional stoneware clays attracted the interest of tea masters from the late sixteenth century. This conscious search for "clay taste" *(tsuchi aji)* would also become an enduring facet of Japanese ceramic style. What is clay taste? While it does not connote taste in the physical sense of the world, it does suggest a distinctive flavor in the clay that is elemental, if not visceral. Therefore, appreciating "clay taste" means savoring the impurities of the clay as an asset rather than treating them as a liability. For example, the buff-gray clays from central Japan (Mino, Seto, Kyoto, Shigaraki) fired to a light color and thus encouraged potters to work with painted decoration and transparent glazes. The yellowish or reddish clays common in western Japan, especially those of Kyushu (Karatsu, Takatori, Agano) frequently fired too dark for painted decoration, so potters used pours of mottled or opaque rice-straw ash glazes. Some clays, like Bizen, were not receptive to glazes at all. But when fired long and carefully, a wonderful range of kiln effects from ash, combustion gases, and stacking enlivened their surfaces.

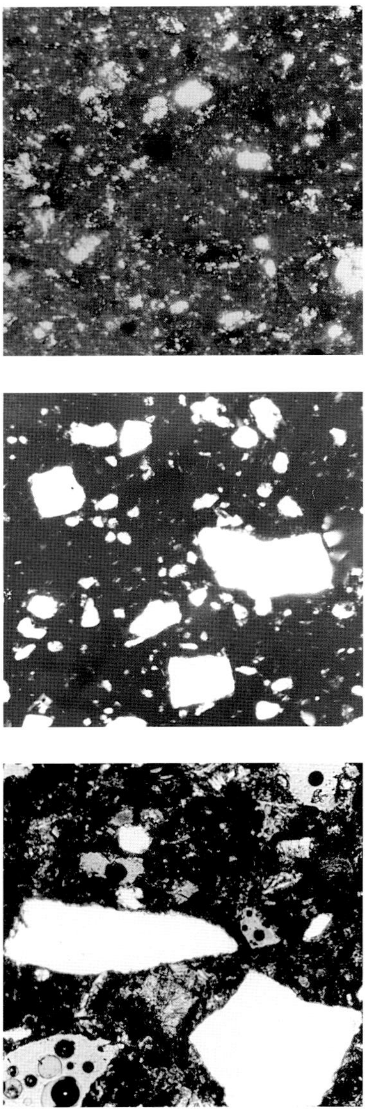

30. Clay micrographs

Technically speaking, clay taste may originate from a number of factors: rough particles of quartz, feldspar, or iron, which may melt or protrude from the surface of the body; or a surfeit of elements such as calcium, which will promote fusion within the body; or alumina, which will work against fusion. Premodern potters had the most control of texture, and its deliberate variation at a single eighteenth-century workshop (that of Kyoto potter Ogata Kenzan) is demonstrated in the 100-power micrographs in figure 30. The example at the top shows an even distribution of fine particles (over 100 mesh); fired surfaces are smooth. The middle example shows an even distribution of fine and medium-size particles (50-100 mesh); fired surfaces are pebbly. The example at the bottom shows an uneven distribution of fine and large particles (10-50 mesh); fired surfaces are relatively smooth, with abrupt breaks on the surface. Firing influences clay taste as well. Even a single clay body will change according to the way it is fired: if it has 4 percent iron, it will turn light red at 800 degrees centigrade, but at 1200 degrees centigrade it will fire to a brown color. It will also change color according to the amount of oxygen in the kiln.

Porcelain, which fires white and achieves a flinty hardness, lends itself to beauty of line or painted decoration. Stoneware clays,

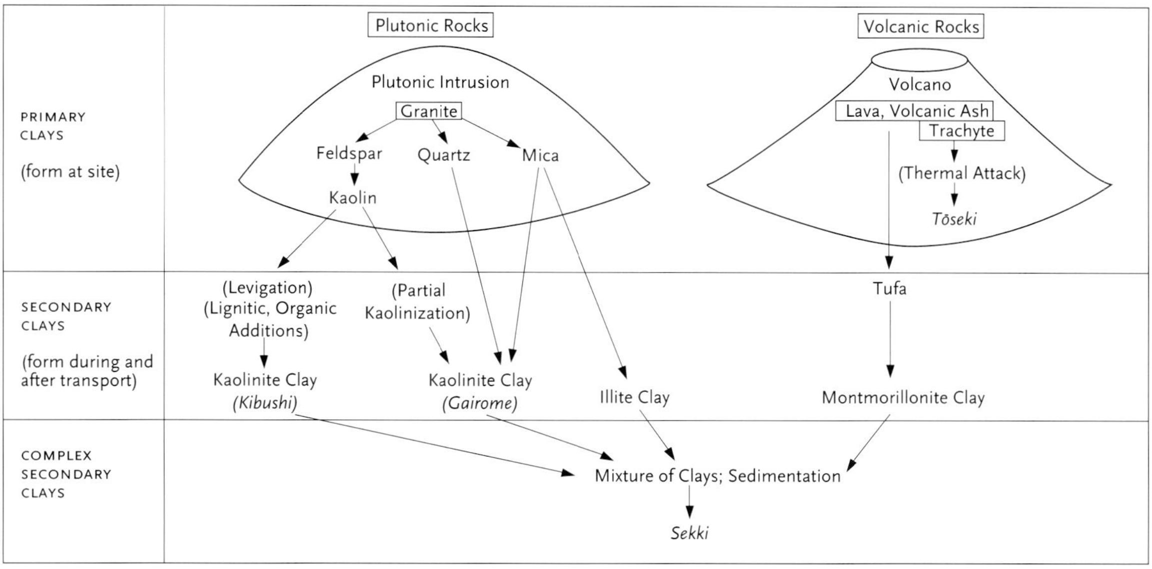

31. Origin-evolution of Japanese clays.

whose impurities respond most conspicuously to firing, are regarded as having the richest flavor. Theirs is a beauty of surface. Earthenwares usually have significant proportions of iron, and this may harmonize with treatments such as burnishing.

Porcelain

All clays contain alumino-silicate crystals that slide over and stick to one another when wet. This material forms from the decomposition of parent rocks, the granites and gneisses that comprise much of the earth's surface (fig. 31). Clays that form at the site of the parent rock, or primary clays, are represented by two major types in Japan, *tōseki* and kaolin. *Tōseki* (literally, "ceramic stone"), the principal component of traditional Japanese porcelain, is similar in composition to the Chinese porcelain stone, petunze. *Tōseki* derives from volcanic rocks such as trachyte and rhyolite, which have metamorphosed under thermal and gaseous attack. The extremely fast cooling of these rocks encourages the formation of a mixture of quartz and ultrafine potash mica (sericite).[2] The fine particle size of the sericite provides plasticity, and the alkali in it also fluxes the body at high temperature. While *tōseki* can be used alone, lesser amounts of ball clay are sometimes added to provide plasticity and dry strength, and feldspar may be added to provide additional flux. The first known Japanese *tōseki* deposit, Izumiyama in Saga Prefecture, and the largest deposit of *tōseki,* Amakusa in Kumamoto Prefecture, have been the traditional supply for the Arita (Imari) porcelain industry. Smaller deposits are found nationwide, usually near hot springs. The closest Western equivalents to *tōseki* are Cornwall stone and Plastic Vitrox clay.

The other primary clay is kaolin (named after Kao Ling mountain in Jiangxi Province, China, where large deposits are found), a by-product of the decomposition of acidic igneous rocks such as granite. Kaolin is an important ingredient in porcelain and white stoneware clay bodies as well as in slips and glazes. For example, the Ibusuki kaolin deposit in Kagoshima Prefecture has supplied the white body for Satsuma ware. Numerous small kaolin deposits are located in Japan, although most kaolin is now imported.

Stoneware

Primary clays typically lack plasticity due to their large particle size. On the other hand, they are quite refractory due to the absence of mineral impurities, which tend to act as fluxes in combination. As primary clays are transported from their original beds through erosion, they undergo a natural sorting of particle sizes and become mixed with other minerals. These minerals affect color, texture, and fusibility. The resulting mixtures, known as secondary clays, have been the chief resource for the stoneware potter, and several categories exist. "Frog-eye" *(gairome)* clay is a class of kaolinic, vitrifiable secondary clay only partially decomposed through transport. Consequently, it tends to include much mineral debris from the parent rock, including quartz, feldspar, and sericite mica. It is most easily identified by its large quartz particles, whose gleam has prompted the analogy to frog's eyes (*gairo* is "frog" in the Seto dialect; *me* is standard Japanese for "eye"). Frequently the clay particles are still large and irregular, with the result that the body lacks plasticity. It is typically low in mineral colorants such as iron. Because of its noduled texture and high alkalinity (high percentages of sodium or potassium, which can contribute to interesting coloration after firing), *gairome* is considered to be the richest in "taste." It is the preferred clay for many tea-ceremony wares, especially glazed stonewares. The Daidō clay of Hagi ware and the moxa clay of Shino ware are famous examples: they are hard to throw and lack plasticity, but they have interesting texture and coloration. These are the clays that potter-prospectors are on the lookout for, and when available commercially they are the most expensive.

"Wood-knot" clay *(kibushi)* is also a vitrifiable secondary clay, primarily kaolinic but also containing other clay minerals such as montmorillonite and illite. Its primary difference from *gairome* is that it has been subject to natural levigation, resulting in a mass of fine clay particles. Organic materials may combine with the clay, making it look dark in its unfired state. Among the deposits are carbonized tree fragments, including the knots that give this clay its name. Due to its smaller particle size and possibly the lubricating effects of organic acids, *kibushi* is generally more plastic than *gairome*. Since it is available in large deposits, *kibushi* is used industrially, but it does not display very interesting kiln effects. An example is Motoyama *kibushi* from Seto. Most formulated stoneware clay bodies now contain both *gairome* and *kibushi* clays, and it is not unusual to find these clays in adjacent layers in the earth.

In addition to *gairome* and *kibushi*, which are comparatively simple secondary clays, there are complex secondary clays that have undergone extensive transport and are often formed from several parent rocks such as shale, trachyte, and tufa. These clays are characterized as *sekki*. Their particle size is usually small, and they are quite plastic. Many of them have been altered by marine sedimentation, and some have high percentages of calcium, iron, or manganese. This cuts down on their refractory properties, and as a whole they tend to be fusible. Many of the famous unglazed stonewares that emerged in the medieval period are of the *sekki* type, such as Tokoname and Bizen ware.

PROSPECTING AND PROCESSING

Prospecting for clay is the privilege of small workshop potters and hobbyists. It is also a way toward a truly local style. It begins with topography. Primary clays and simple secondary clays are usually found in mountains and hills, while more complex fusible clays tend to be found in alluvial flatlands. Construction sites, river banks,

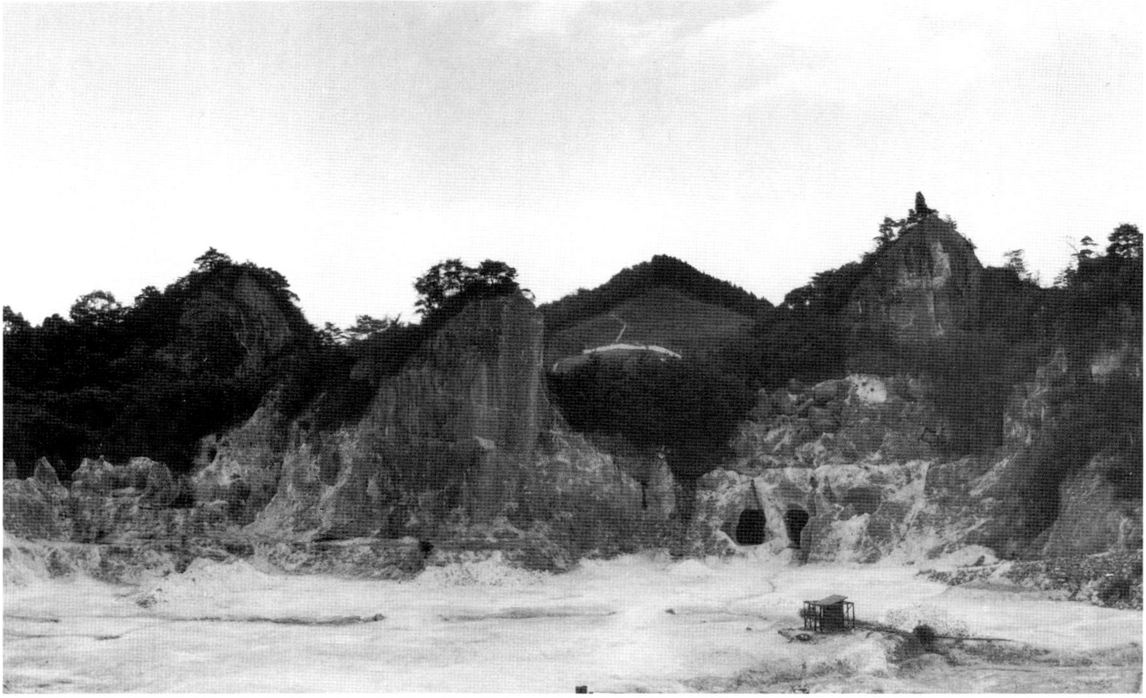

32. Izumiyama clay mine, Arita

and road cuts are the easiest places to find clay. Veteran potters in Japan have various rules of thumb about prospecting: good clay can be found under bamboo groves and ponds, especially in the mountains; craggy mountains yield primary clays and stones—Izumiyama at Arita (fig. 32) is an example—and rounded mountains yield secondary clays. Clay from northern exposures is not well decomposed; clay from southern exposures is more decomposed. When clay is found, it is tasted and smelled. If it is bitter, sour, or rotten it is either too alkaline or too high in organic materials. Newly discovered clay is immersed in water. If it dissolves quickly, there is a high percentage of stony material such as quartz. If it dissolves slowly, there is a high percentage of pure clay. In every case, the best way to test a clay is to make something out of it and fire it to a high temperature. Attention is paid to wet and dry strength, shrinkage, deformation in the kiln, and compatibility with glaze. Clays that lack plasticity will need plastic clay such as a *kibushi* or Western ball clay. Clays that shrink too much will need a nonplastic filler such as sand or grog. The high-iron, high-shrinkage fusible clays of the *sekki* type do not take glaze well, and it is no coincidence that Japanese potters have traditionally left these unglazed, sometimes burnishing them. It is important to realize that there is potential for nearly every clay, and the challenge is to find the application that maximizes it.

What do contemporary potters do when they have access to only pasty commercial clay bodies? Traditionally, Western potters have covered their pots with glazes that obscure the foundation material. They have never consciously exploited a visible interaction between glaze and body. As long as glaze is treated like paint, as it commonly is in the West, it is inevitable that the qualities of clay will remain ignored. In Japan, however, mutual enhancement of materials and methods has long been a priority for potters. Some of these historic examples are discussed in Chapter 7.

For potters who have no access to usable raw clays, the Kyoto approach to clay is instructive. Kyoto, a great cultural center, was short on materials but long on ideas about how to make refined versions of other Japanese wares. This should have required a great variety of clay bodies, but the Kyoto potters learned to do this with a few generic ingredients. As a base they used the buff stoneware from nearby Shigaraki, which was cheap and plentiful. To make a fine white stoneware resembling their own Kurodani clay, which was already depleted in the eighteenth century, they added roughly 20 percent kaolin to the Shigaraki base. For *gohonde* clay, a body that sensitively reacted to changes in the kiln atmosphere, they added 20 percent fireclay to the above Kurodani substitute, so that large particles in the fireclay created glaze pinholes, around which pink halos formed when the kiln reoxidized. To imitate Karatsu clay, a richly noduled body from Kyushu, they added to the Shigaraki base 20 to 30 percent quartz sand, graduated in size between 60 and 100 mesh, and increments of iron-bearing clay. For a facsimile moxa clay, the much-admired stoneware body used in Shino wares, they added increments of the Shigaraki base clay to kaolin until the body was manageable. For *nambande,* a rough red body that emulated early unglazed tea-ceremony wares, they added 40 percent ochre and 50 percent powdered charcoal to a coarse Shigaraki base clay. As demand for certain styles exceeded the original clay supplies at other kiln centers, the Kyoto blending approach became standard practice, and today the name of a clay now attests more to its aesthetic performance than to its geographic origins.

The Kyoto potters also knew that preparation of the clay would affect its appearance. There are three basic stages in clay processing, and depending on the desired effect and the clay at hand, a usable

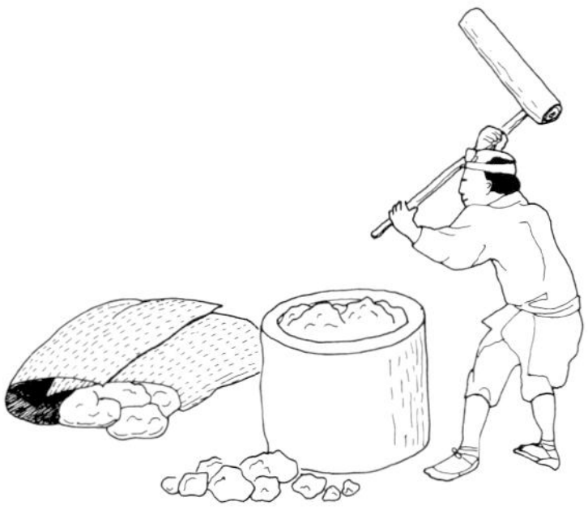

33. Clay pounding

clay body may be obtained at each stage. The first is direct use: the clay is used "as is," with stone and organic impurities left in or removed by hand. If possible, the clay is aged before use. As it breaks down to its finest particle size, lubricating films form around individual rather than aggregates of clay particles, making the whole mass more plastic. Alternate drying and wetting tend to speed the breakdown (similarly, potters in northern Japan remark that freezing and thawing make their clays more plastic), so rather than covering the clay with vinyl it is preferable to expose it to air and wet it from time to time.

The second stage is dry processing. The clay is dried well and pounded. Before the days of power crushers and ball mills, clays were pulverized with wooden mallets (fig. 33) or in water-driven mortars (fig. 34), some of which can still be seen in country potteries. The old method is said to result in flatter clay particles, which

34. Water crushers

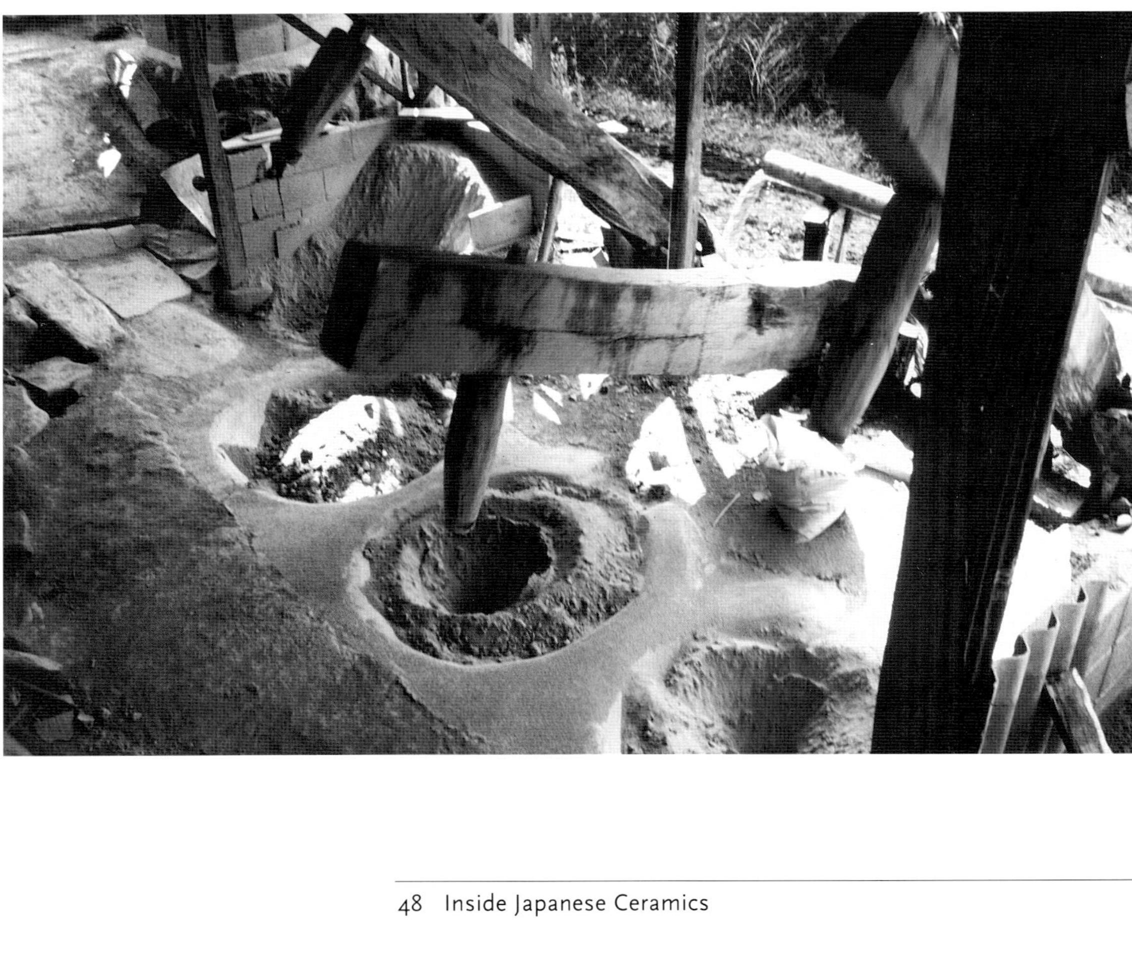

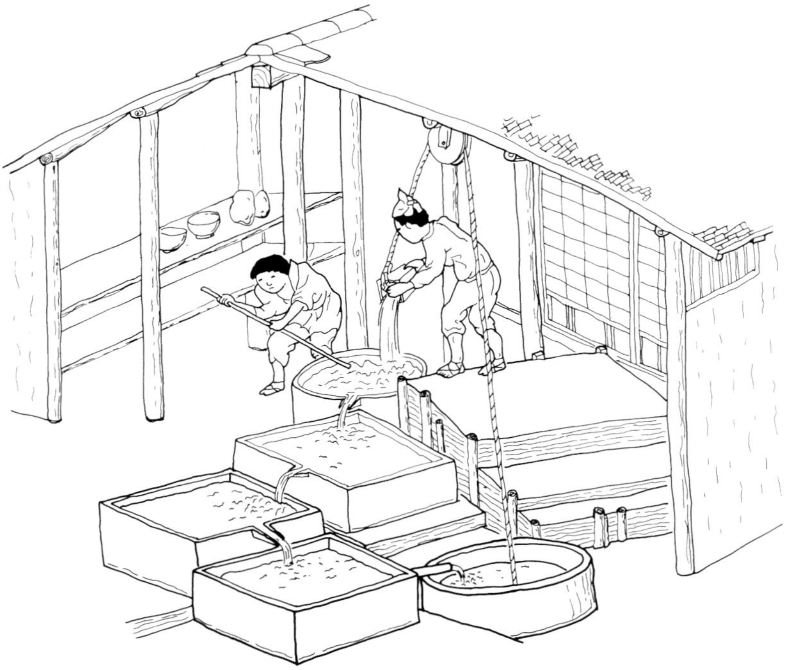

35. Levigation

have more plasticity than the round ones that come from machines. For use at this stage, the dry, pulverized clay is sieved through a heavy-duty screen (30–60 mesh). It is then wetted and kneaded. This type of preparation is frequently used for the "tasty" *gairome* clays from Hagi, Karatsu, and Mino.

The third stage, which was first consistently employed in the seventeenth century, is wet processing or levigation (fig. 35). The clay is dried, crushed, and immersed in water so that the heavy material sinks to the bottom. The finer particles remain in a suspension, which is transferred to an adjacent lower tank. As this slurry is moved from tank to tank, progressively finer grades are obtained. Since clays from alluvial deposits have already undergone natural levigation, one simple step suffices to divide coarse, intrusive material from fine particles. Levigated clays are dried in their settling tanks or ladled out to dry on straw mats, in unglazed bowls, on roof tiles, or even on small kilns built for this purpose. If a clay is found to lack plasticity, a more plastic clay such as *kibushi* is added. In the past, some potters added salt, probably for clays with a high percentage of lime, which tends to stiffen a clay body. Rice husks were another additive. As they decomposed, they contributed organic acids that increased plasticity. Handmade paper *(washi)* was occasionally repulped and added to enhance plasticity and dry strength.

Kneading, the first hands-on chore of the pottery apprentice, is necessary to even out the hard and soft spots in clay bodies and to eliminate air pockets. There is less warping and cracking in a carefully kneaded clay body, and kneading also helps to develop plasticity. Even potters who use a pugmill maintain that hand kneading imparts these qualities.

When the clay is moist but too hard to knead, it is cut into fist-size lumps or pancake slices and immersed in water for about five minutes. The lumps are withdrawn and stored in plastic for a few hours before kneading.

Foot-kneading (fig. 36) is a simple way to homogenize and remove the air from a large mass of levigated clay. Although rarely seen in Japan nowadays, this technique is still occasionally used in Korea and other Asian countries. The potter steadies the upper body by hanging on to a wooden crossbar or rope suspended from the ceiling and treads on a mound of wet clay with the heel or side of the foot. The idea is to work around the circumference of the mound, forming a spiral pattern that radiates from the center. When the clay mound is fully distributed into the spiral, it may be scraped up for hand kneading.

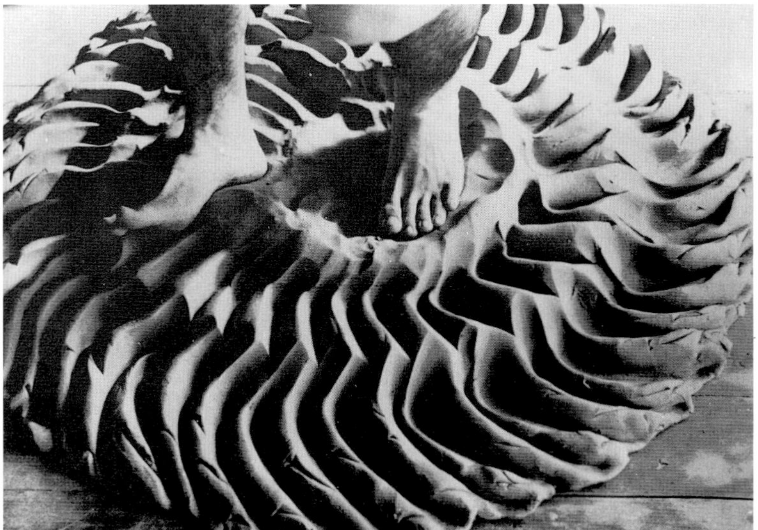

36. Foot kneading

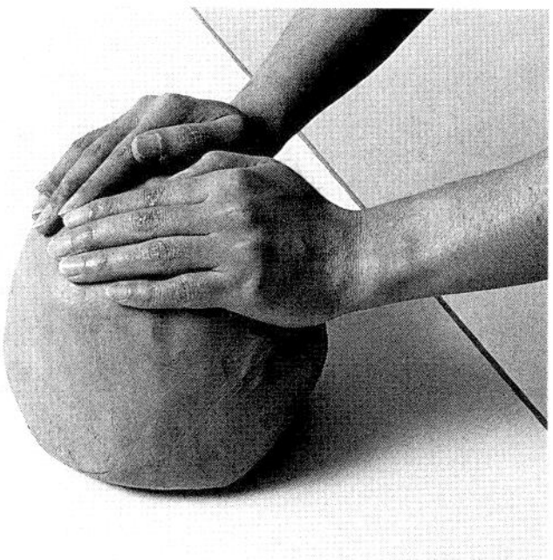

1

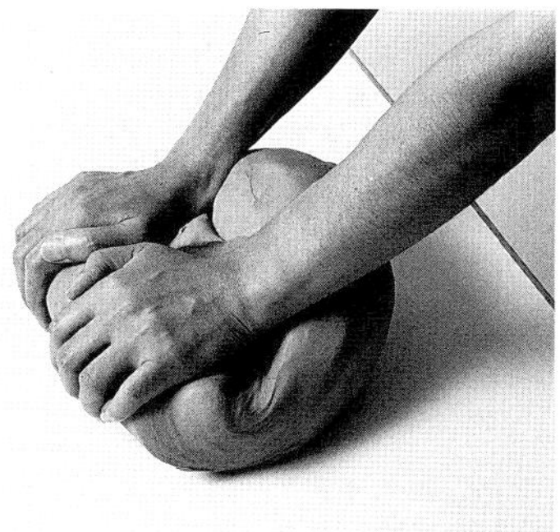

2

37. Rough kneading

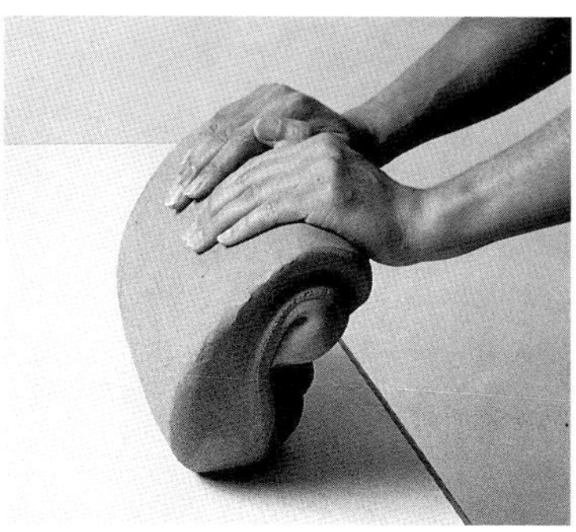

3

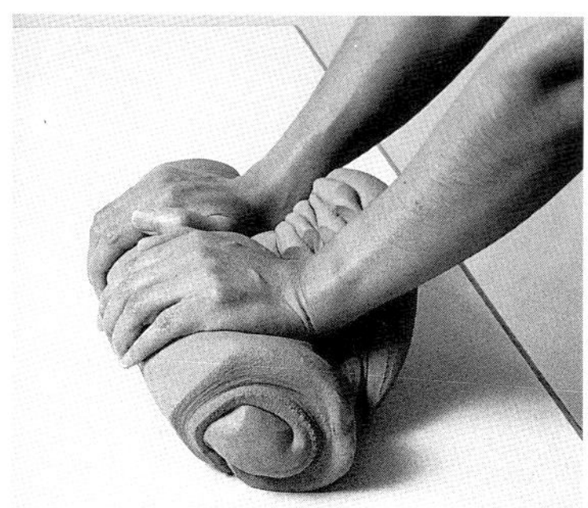

4

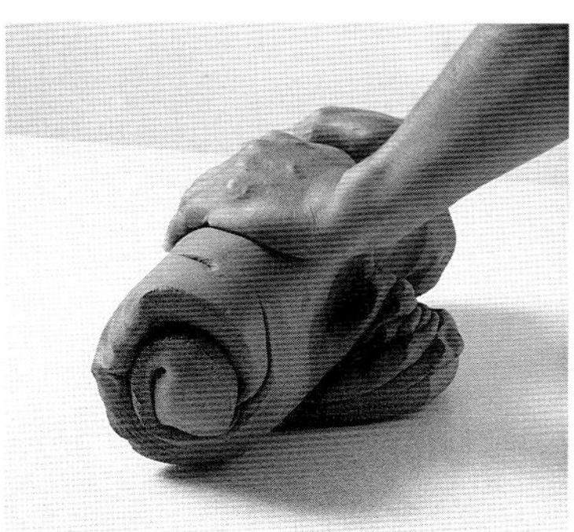

5

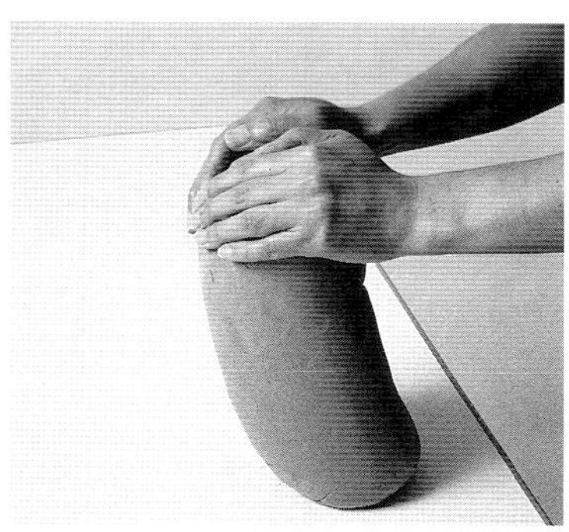

6

Rough kneading (fig. 37), sometimes called "ram's-horn" kneading in the West, is used to homogenize the clay mass after an initial removal of air by foot or machine. A mass of clay, averaging 3 to 5 kilograms, is patted into an oblong shape on a dry kneading table. The hands are placed next to one another, palms down, on top of the clay (1). With force concentrated in the bottom of the palms, the clay is pressed down and away from the body (2). The front end of the flattened clay is grabbed, the mass is peeled back up, and the hands move forward on it (3). The down-and-away press is repeated (4). After about ten presses, the clay will have spread out to both sides (5). Both sides are folded inward into a single lump, and the pressing sequence is repeated once more (6). The rate of homogenization may be understood by starting with a lump made of two differently colored clays: when the original colors are indistinguishable, the rough kneading is finished.

Spiral or "chrysanthemum" kneading (fig. 38) is especially effective in driving air holes, even small ones, out of the clay mass. Prior to my first trip to study ceramics in Japan, I received two blurry snapshots of this process from my master-to-be with the terse instruction "learn this." Learning it takes repeated practice and patience, but it is a way of finding out about the day's clay, which is never the same as yesterday's. These instructions are for the counterclockwise-turning wheel. (For the clockwise wheel, hand positions and turning direction are simply reversed). Basically the right hand does most of the work, pressing down the clay while rolling it forward. The left hand acts as a guide, lifting and pivoting the mass after every push. It is difficult to do well.

To begin spiral kneading, the hands are placed on the top of a lump of clay, about 8 kilograms at the most (1). The clay is pressed down and away from the body (2). Most of the force, about two-thirds, is channeled through the right hand. The force must come from the entire body rather than from the arms and wrists, or else one tires very quickly. Then the clay is lifted for the next press. As the left hand lifts and simultaneously rotates the clay clockwise, the right hand moves forward, around the "back" of the mass, establishing a new grip (3). The clay is pressed down and away again (4). As this is repeated, a series of folds radiates from the right of the conical clay mass. These folds are the "petals" of the chrysanthemum. As the pressing and turning continue (5), the petals are folded back into the mass (6). After about 100-150 of these presses, the clay is consolidated gradually into a ball. With the left hand continuing to pivot the clay from the butt-end, the right hand moves closer to the tip of the cone with each press. The "flower" decreases in size (7). The clay forms into a bullet shape, with the last flower petals being wrapped around its tip (8). The clay is slapped into a final shape suitable for throwing (9).

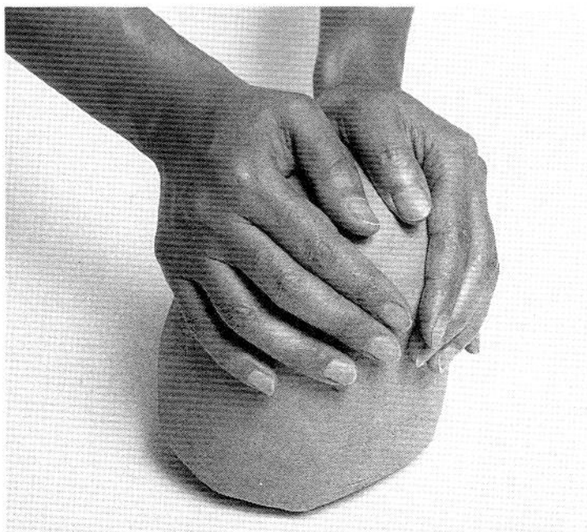

1

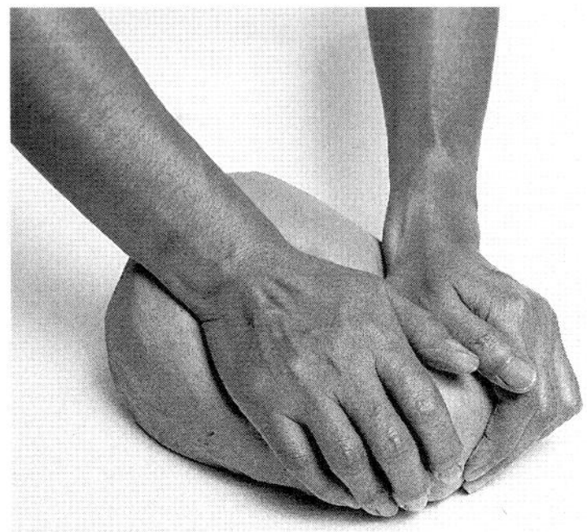

2

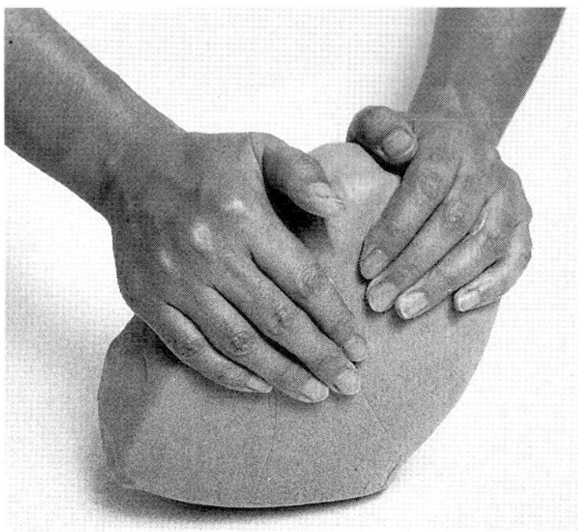

3

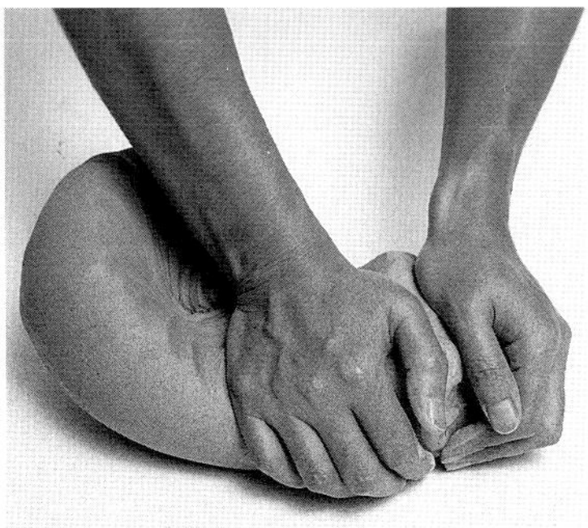

4

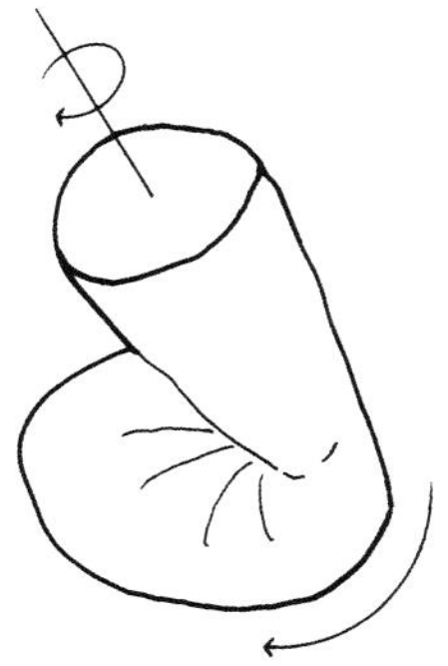

5

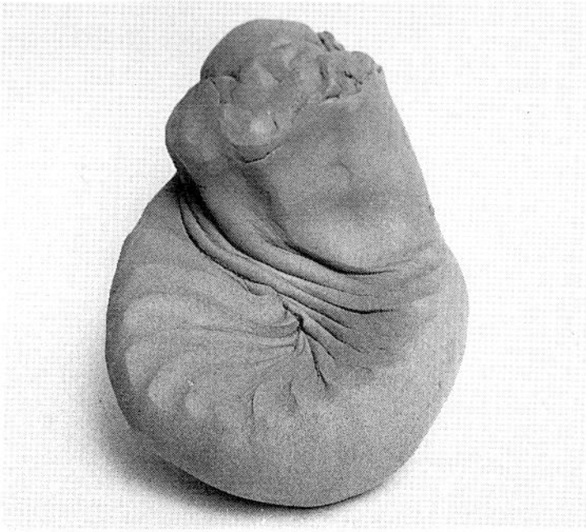

6

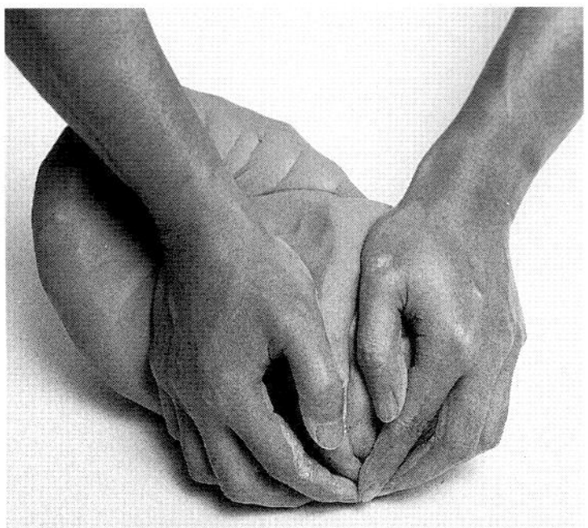

7

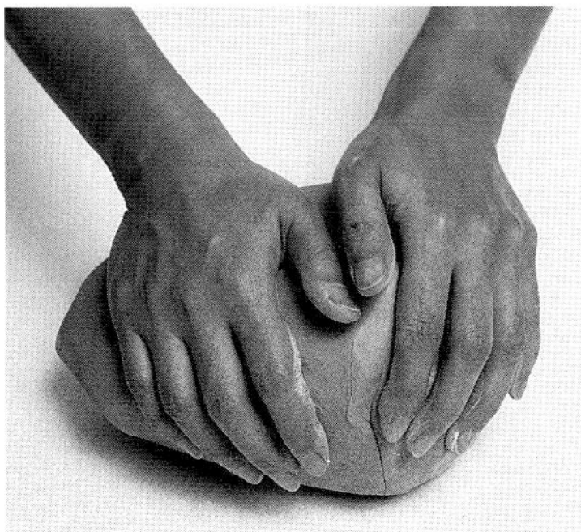

8

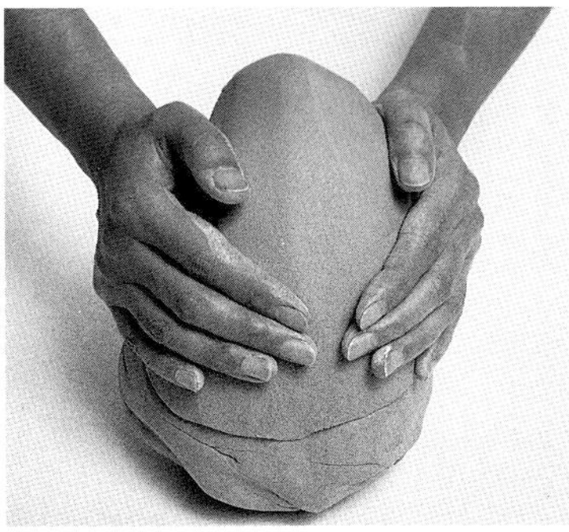

9

38. Spiral kneading

In the days before pugmills, the clay was generally foot-kneaded and rough kneaded, and then spiral kneaded 60 times from each end, twice. The clay is usually kneaded just prior to use, for unevenness from drying can occur quite quickly. Now ready for use, the kneaded clay balls are covered with plastic and kept off absorbent surfaces.

NOTES

1. Brinckmann, Justus. *Kenzan: Beiträge zur Geschichte der japanischen Töpferkunst* [Kenzan: Commentaries on the history of the Japanese ceramic art] (Hamburg: Lütcke and Wulff, 1897).

2. Wood, Nigel. *Oriental Glazes* (London: Pitman Publishing, 1978).

THREE

Forming

Even in the prehistoric period, Japanese potters employed a diversity of forming techniques. Coiling was a principal method of the Neolithic Jōmon potters, the coil-and-paddle method appeared at the advent of the common era, and just a century or so later there is also evidence of finishing, although not forming, on a simple turntable. Slab-forming and drape-molding techniques appeared from about the fourth century. The first use of the potter's wheel in Japan can be seen in Sue ware, which emerged in the late fourth or early fifth century under the influence of Korean stonewares. While the Sue wares combined wheel and coiling (coil-and-throw) techniques, the lead-glazed earthenwares made from the eighth through tenth centuries included forms made entirely on the potter's wheel. Combinations of the above persisted into the early modern period. By that time, most potters in central Japan were working with the large hand-turned wheel, which was probably introduced from China in the fourteenth century, and potters in western Japan were using the fast-turning kick wheel, imported from Korea in the late 1500s.

From the mid-nineteenth century, newly introduced Western techniques revolutionized the way the Japanese made pots. Today, electric jigger-wheel, slip-casting, and vacuum-press processes make it possible to duplicate forms with precision and efficiency. But industrialization robs ceramics of intimacy: many ceramics factories are indistinguishable from those making plastic. Despite the technological conveniences, age-old forming methods continue to be used, supported by an entire nation of patrons who find warmth and spontaneity in handmade ware.

HAND TECHNIQUES

The traditional techniques include pinching, coil making and building, carving out, slab making and assembly, drape molding, press molding, and piece molding.

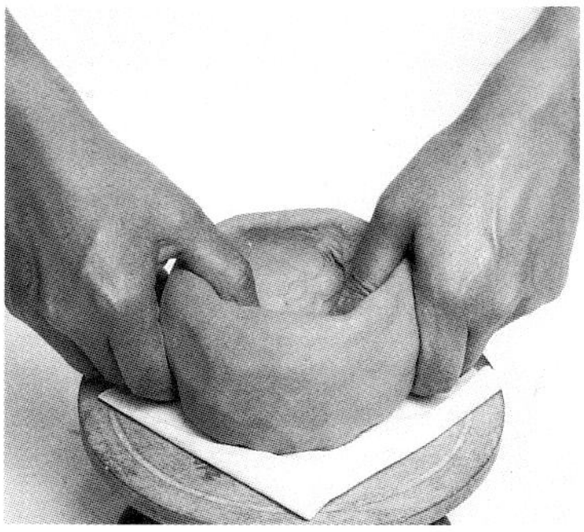
1

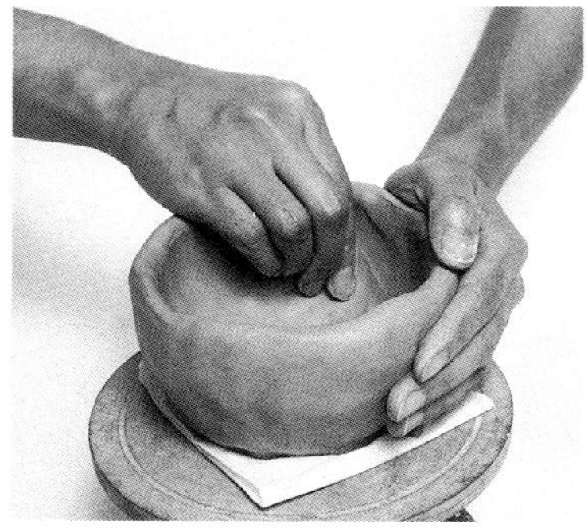
2

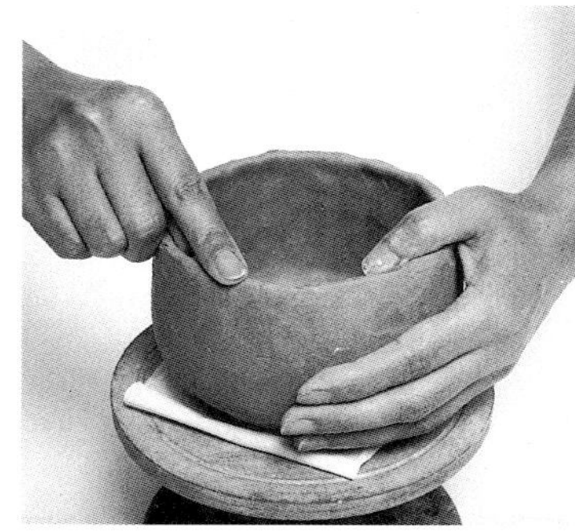
3

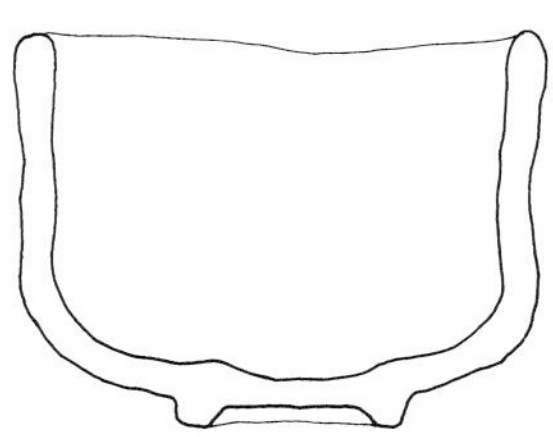
4

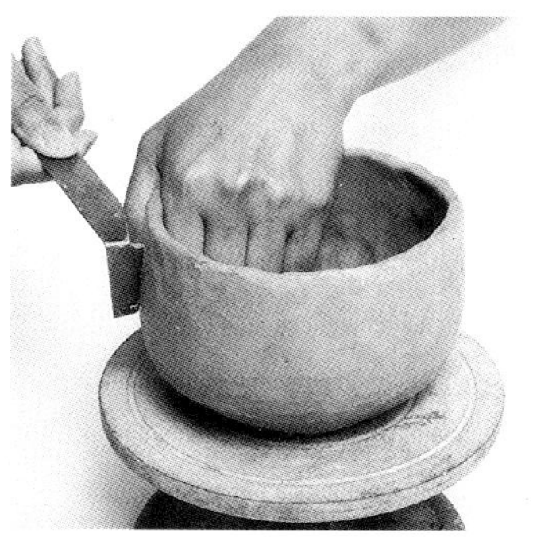
5

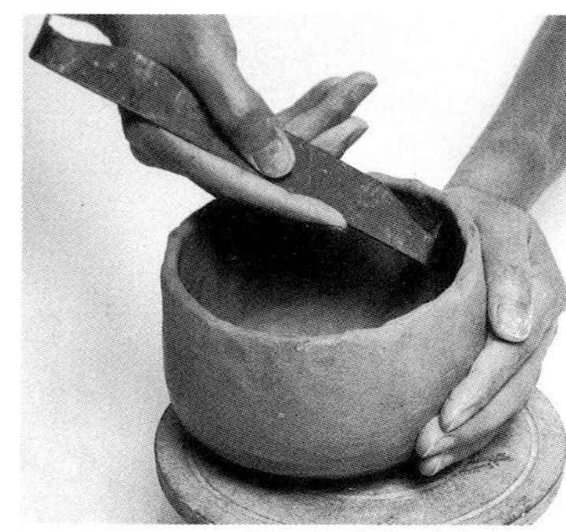
6

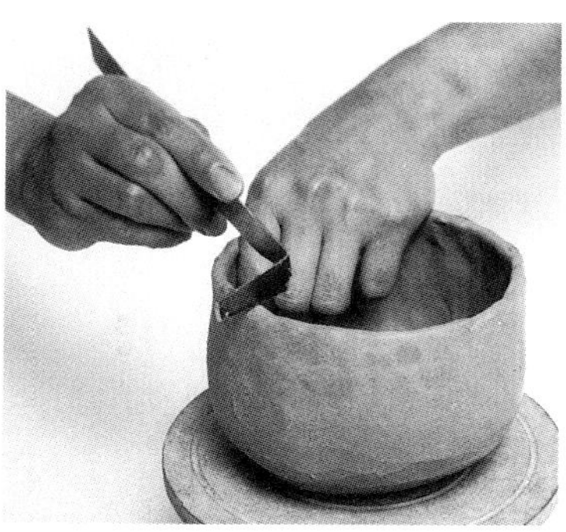
7

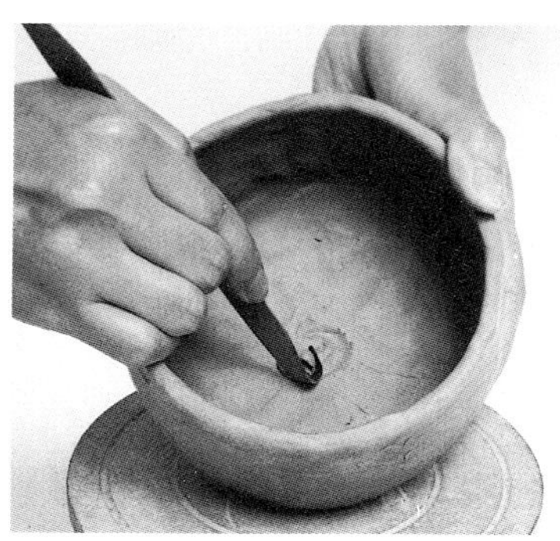
8

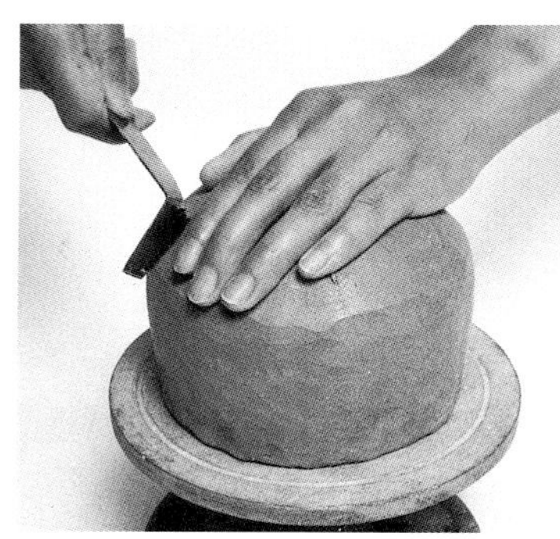
9

39. Pinching/carving

Pinching

Because of its simplicity and historical associations, pinching occupies a special place among Japanese forming techniques. It is one of the preferred methods for the manufacture of tea-ceremony bowls, for users appreciate the irregularities and rich tactile sensations of hand-formed pots. No matter how skilled one may be with the wheel or

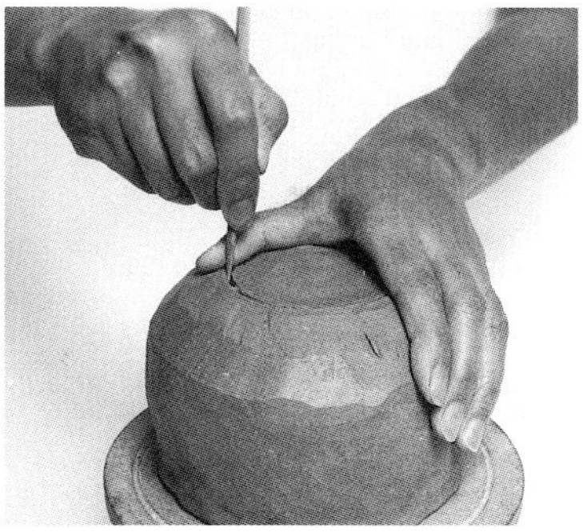

10

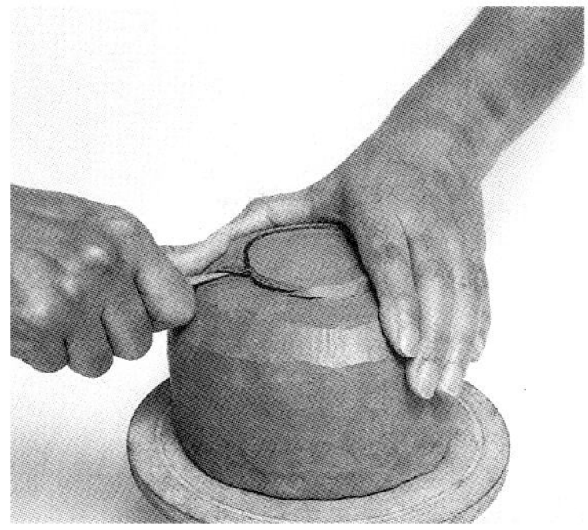

11

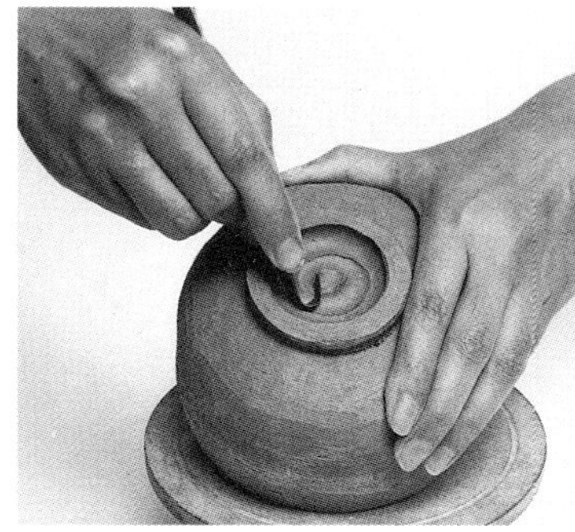

12

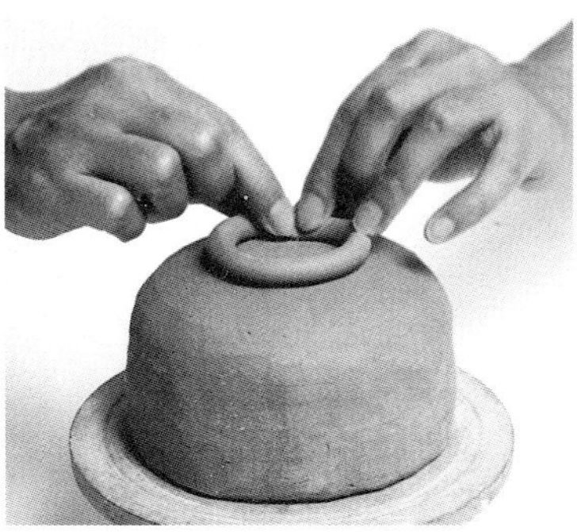

13

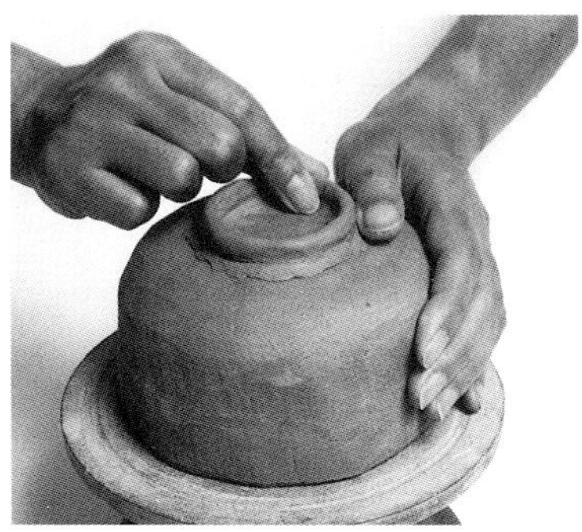

14

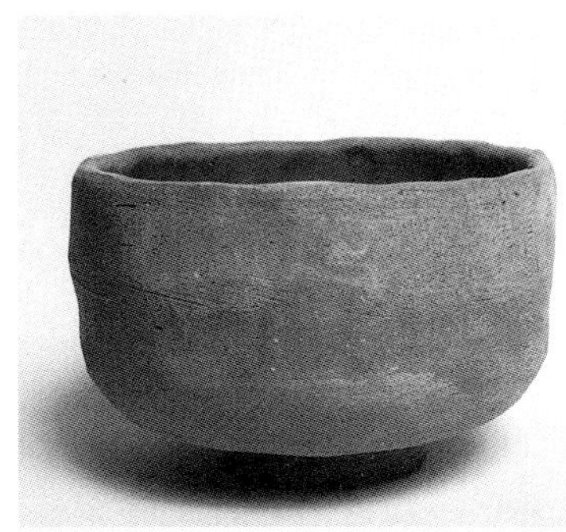

15

molds, it is a challenge to pinch a thin, even-walled pot. Pinching is also a way to become acquainted or reacquainted with the fundamentals of form and the manner in which parts of a pot relate to one another.

Pinching a small bowl (fig. 39). A round ball of clay about the size of a grapefruit is fashioned and either held in one hand or placed on a turntable. One or two thumbs are inserted to form a round hollow (1). With the one hand turning the ball, the clay is pinched between thumb and forefinger (2). The rim is smoothed (3). The trick is to pinch evenly and to maintain the desired shape. It is important to work at a deliberate pace, but if too much time is taken, small fissures will develop in the surface of the pot. This is a sign that the clay is losing its plasticity.

To achieve a conventional tea bowl shape (4), the bowl is trimmed after it dries to leather hardness. The outside walls are trimmed first (5), and then the inside (6). The rim is beveled (7). Traditionally a small spiral depression is carved into the interior bottom (8). With

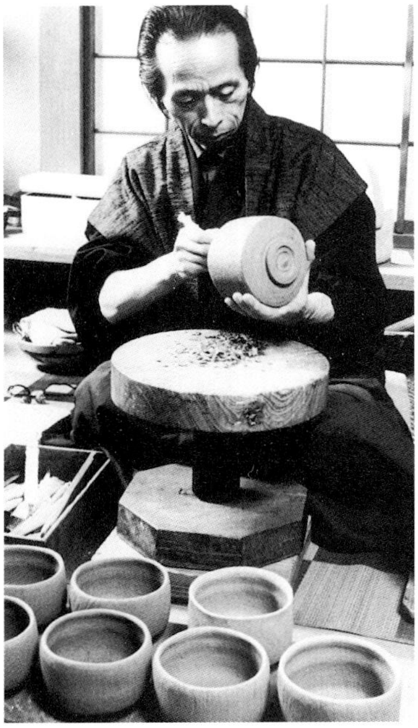

40. Ōhi Chōzaemon carving Raku bowls

care taken not to damage the mouth rim, the bowl is turned over and the base is beveled (9). The outline of the foot ring is traced while the bowl revolves on a turntable (10). Excess clay is trimmed from around the foot ring (11). Finally, the inside of the foot ring is trimmed out, most commonly in a spiral-like pattern (12). A foot ring may also be added by attaching a coil of clay (13). The outside seam is smoothed first, followed by the inside (14). With this, the bowl is finished (15). A good foot ring should give physical stability to the pot and make a conclusive visual statement.

The pinching-carving combination is most frequently used by Raku-ware potters (fig. 40).

Making Coils

Making pots by stacking coils of clay is one of the oldest Japanese forming techniques. Jōmon ware and the celebrated *haniwa* tomb sculptures, for example, were made with either coils or coil-like slabs of clay. Coiling allows greater freedom to improvise than wheel or mold work, since one can significantly alter the form in mid-course. Since it is not a difficult technique, coiling is also practiced by beginners.

Japanese potters make the coils between their hands rather than rolling them out on a flat surface (fig. 41). A small piece of clay is fashioned into a conical shape (1). The cone is inverted between the hands and rolled back and forth (2). Pressure is exerted from the "blades" of the hands against the tapered part of the cone, creating a coil of clay (3). It takes some practice to make an even coil, but once the method is learned, it is much faster than rolling out a coil on a

41. Coil making

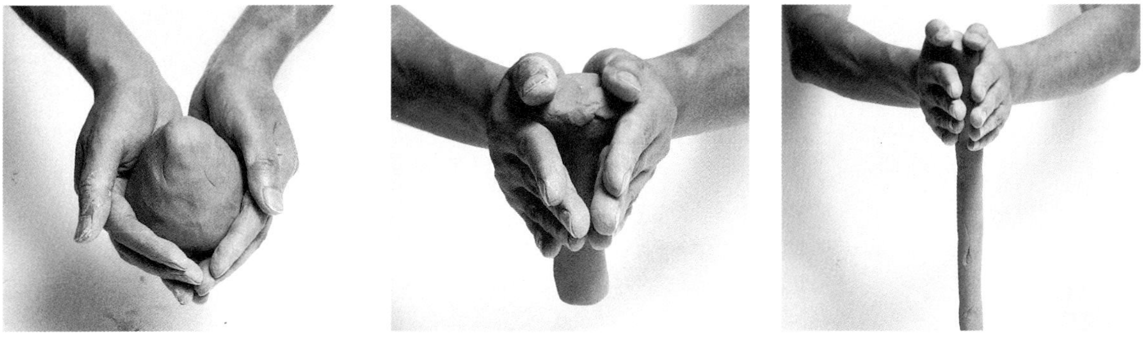

1 2 3

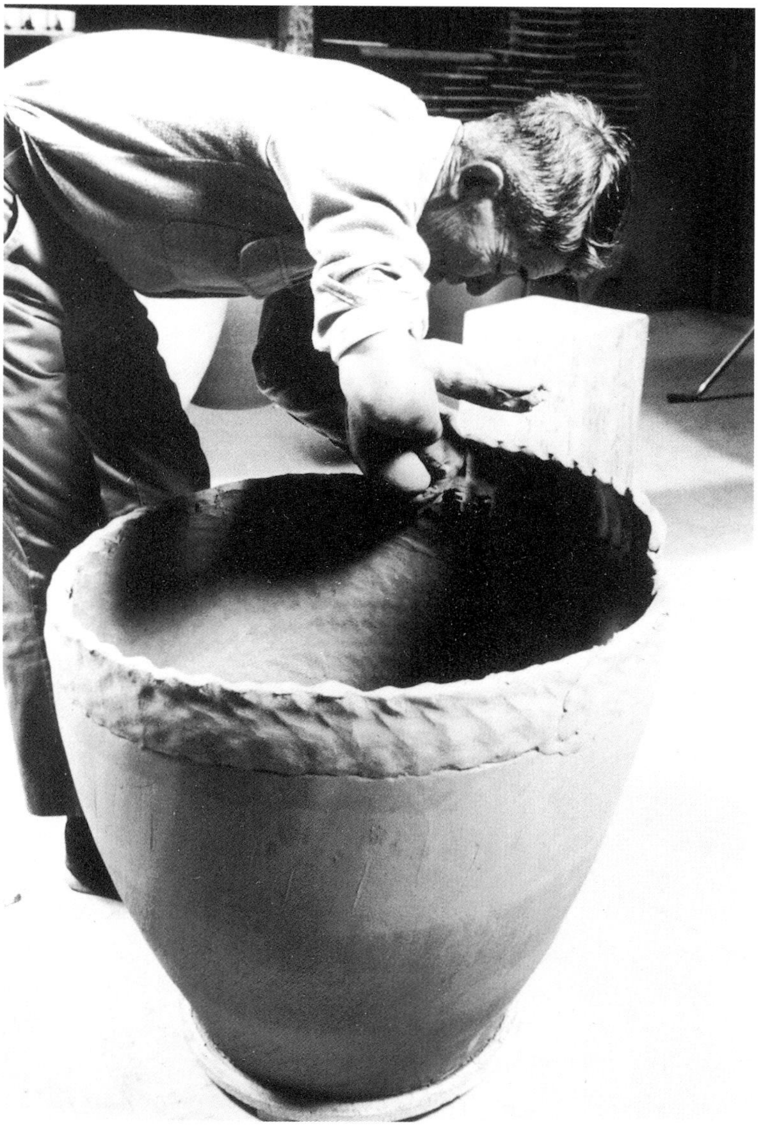

42. Fujita Jūrōemon using the nejitate technique

flat surface. The coils are stronger, too. The coil thickness naturally varies according to the size of the vessel.

Coil Building

Fusing the coils together is the critical part of coil building, and there are various ways to accomplish this. One of the most common techniques, called coil-and-paddle, has been used since prehistoric times, especially for large objects such as storage jars. Ringed coils are stacked into a cylinder, and blows by a wooden paddle seal the joints. A small wooden anvil held along the inner wall of the vessel cushions

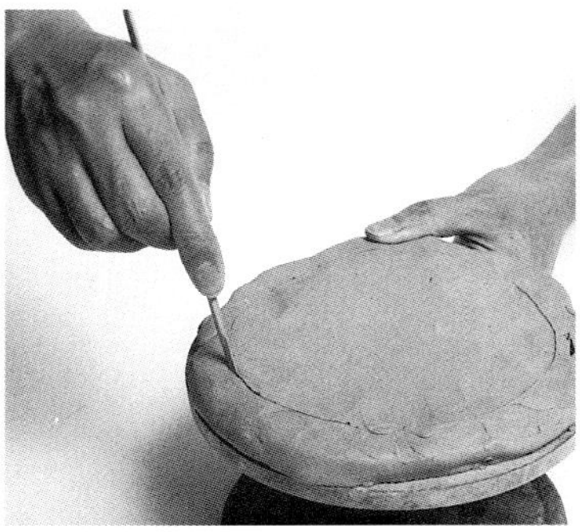

1

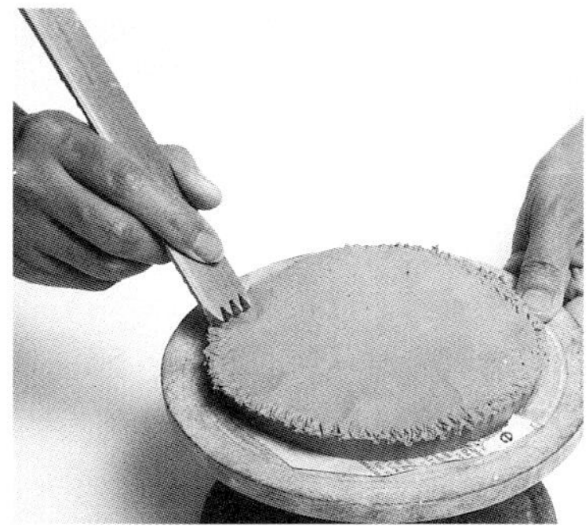

2

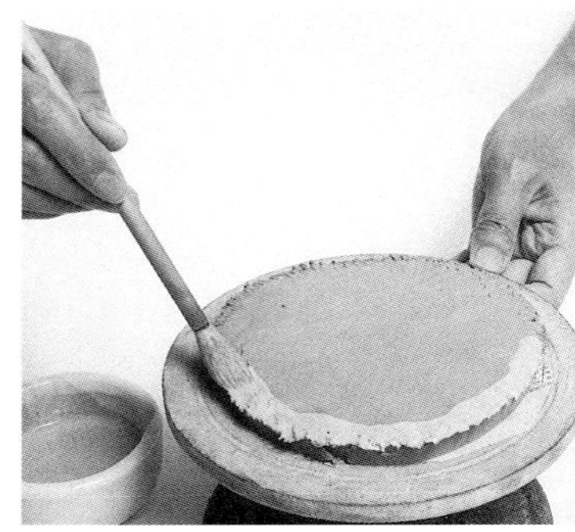

3

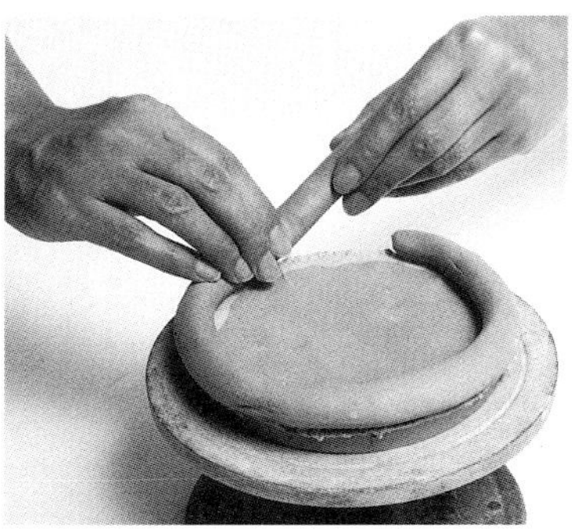

4

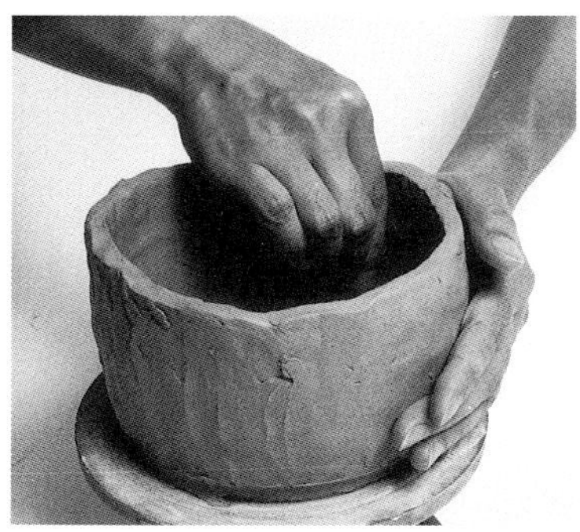

5

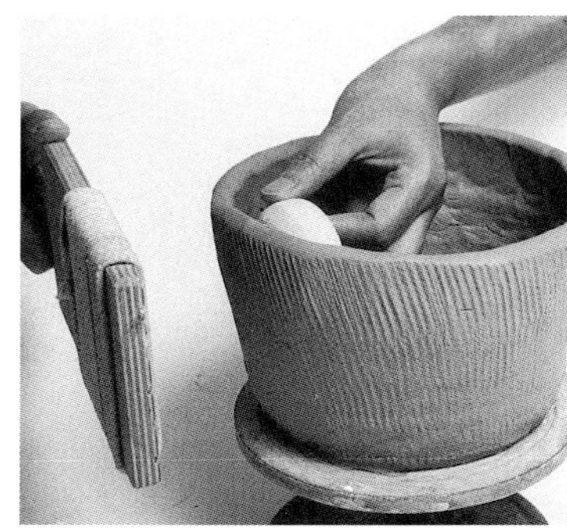

6

43. Coil-and-paddled jar

the blows. Since a smooth surfaced paddle sticks to the clay, paddles are usually textured, and this may add an element of decoration. To produce a large vessel, this coiling and paddling is done in a number of phases. The potter allows the lower portion to dry and support the work above it. Sometimes an incandescent bulb or a small charcoal brazier is suspended inside the pot to speed the drying.

A common technique from medieval potteries, and one still practiced today by a few potters, is called *nejitate*: as each coil is added, the potter simultaneously compresses and draws up the wall when "walking" around the pot (fig. 42). Many potters also combine the coil-and-paddle technique with throwing (coil-and-throw).

Coil-and-paddled jar (fig. 43). The bottom of the vessel is started by paddling out a disk of clay on a turntable. The disk is compressed with the hand or an anvil, starting from the center and moving outward. With the turntable spinning, a circle is drawn to define the

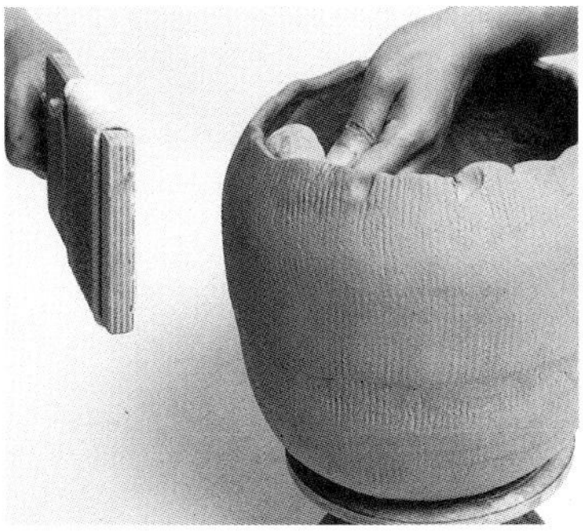

7

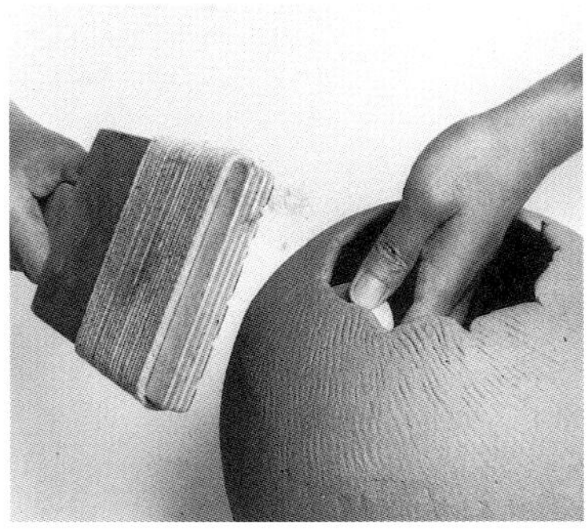

8

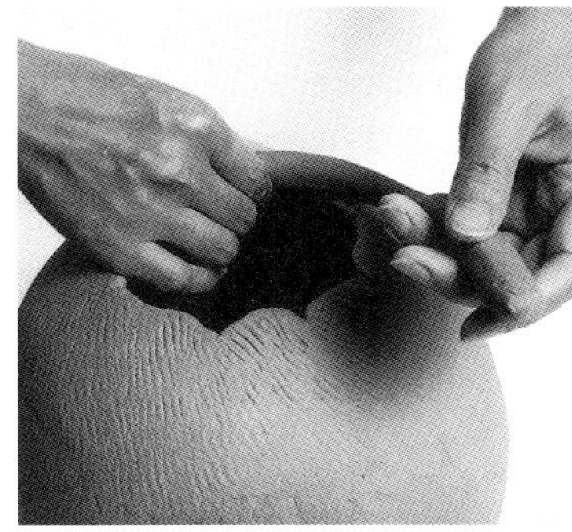

9

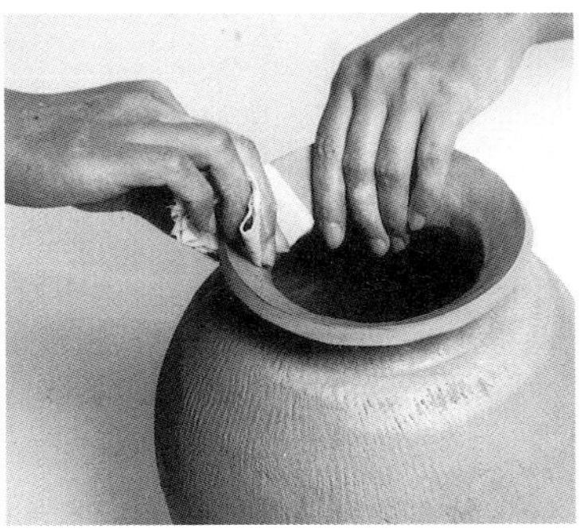

10

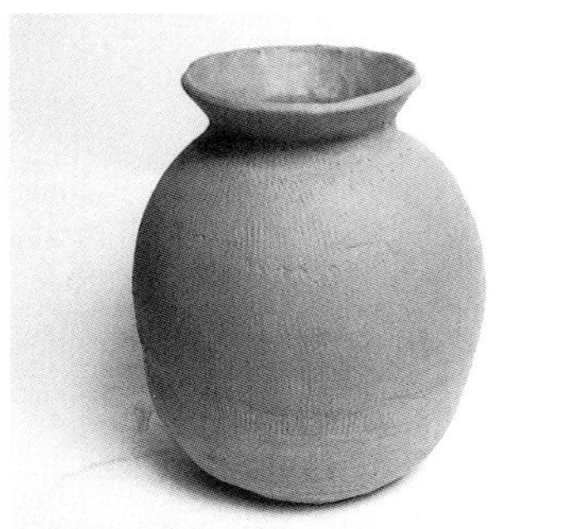

11

base (1) and the excess is cut away. The perimeter is cross-hatched (2), and slip is applied to that area (3). The first coil is laid on the perimeter (4), and then a succession of coils are stacked over that. After about five coils are in place, the seams are smoothed together. With one hand supporting the outside wall, the inside coils are fused with upward pulls of the fingers of the other hand (5). The outside is then paddled; the other hand holds an anvil along the inner wall (6). The wheel is turned slowly during the paddling, and care is taken to maintain the intended contour. The wall rises and becomes thinner (7). Further coils are added and paddled as needed. In forming the shoulders of the vessel, the paddle is tilted inward when striking (8). A neck is formed by adding coils (9). After the neck and mouth rim are formed, a moistened chamois is applied (10). In the process of building up the pot, the bottom will become dry, and excess clay can be trimmed away to a finished profile. The pot is removed from the turntable and set aside to dry (11).

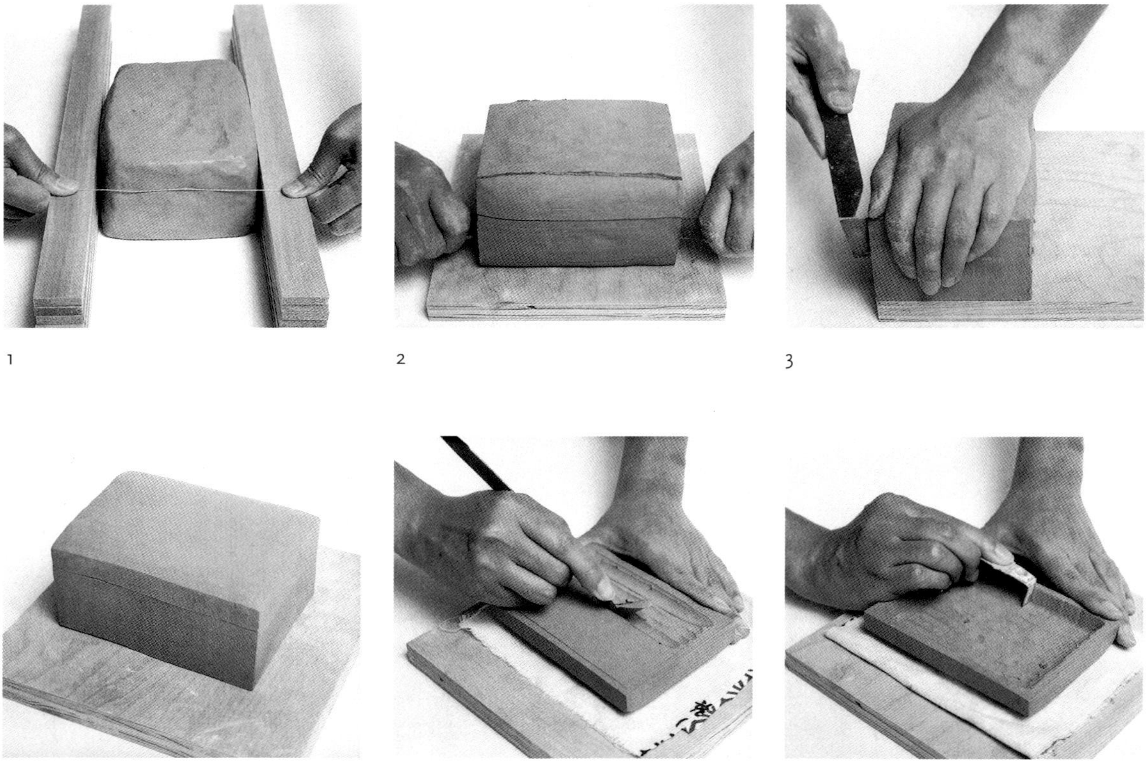

44. Carving out

Carving Out

Here the finished piece is completely carved out of a solid block of clay. This is an especially appropriate technique for lidded boxes or any pieces that must fit together. Compared to pieces made on the wheel or fabricated from slabs, there is less distortion in the drying process. The late Tomimoto Kenkichi produced his well-known large lidded porcelain boxes by this technique. With coarser clays, carving out may also be used to produce an organic, sculptural effect.

Carved-out lidded box (fig. 44). A well-kneaded clay mass is sized up, with allowances made for carving and shrinkage. A block about twice the size of the finished piece is necessary. Since the lid will overlap a flange cut out of the base, the height of the flange must be added to the height of the initial block. A horizontal cut is made by pulling the taut cutting wire through the clay, using the measuring sticks as guides (1). This results in two pieces, a lid and a base. The piece is dried, with lid in place, until it is leather hard. Excess clay is

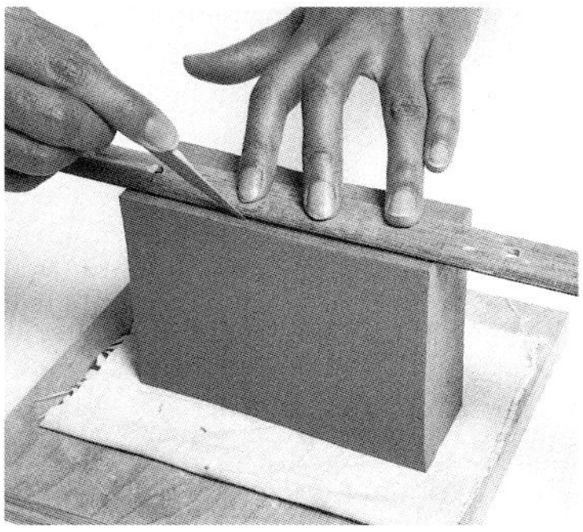

7

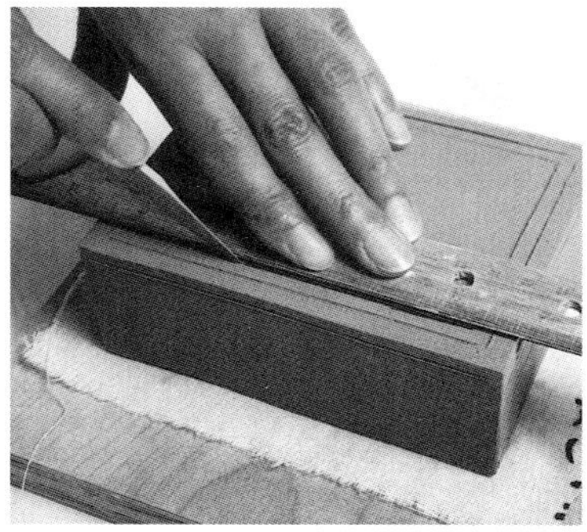

8

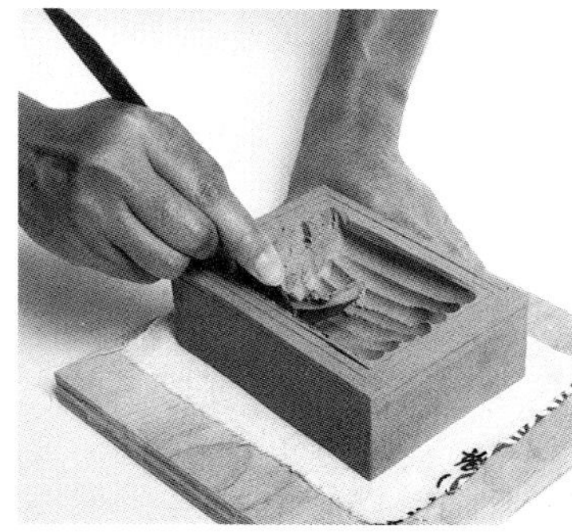

9

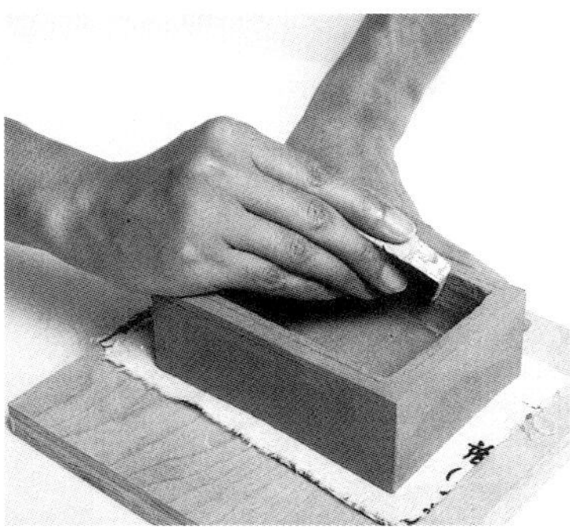

10

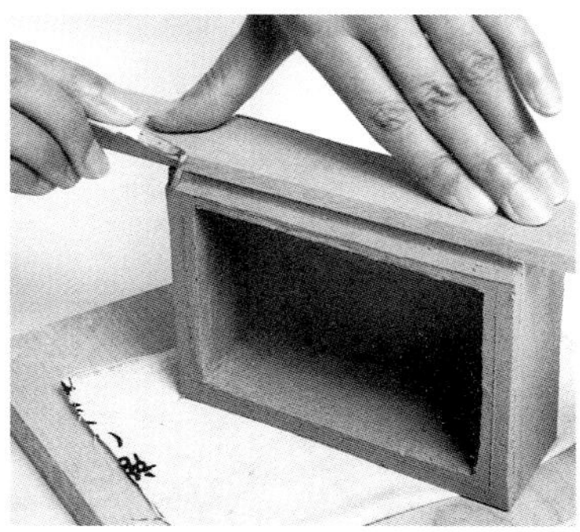

11

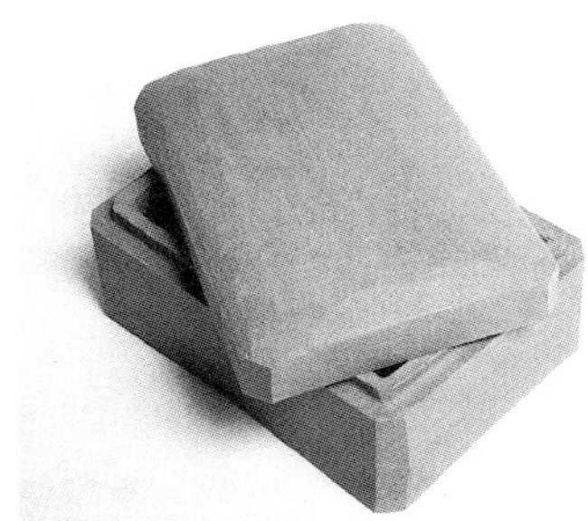

12

sliced away from the sides with a knife (2). With the lid held firmly in place, each side is roughly carved with a knife or serrated-edge tool (a hacksaw blade works well here). The edge of a bat is used as a guide for the final carving of the walls (3). The top of the lid is then carved and left slightly domed so it will not become concave during the firing (4). With the lid resting on a soft cloth, its inside is carved, leaving walls about 5 millimeters thick (5). The carving is finished with a sharp tool (6).

To determine the lower edge of the flange, a ruled line about 4 millimeters below the upper edge is incised on one side of the base (7). To continue that line evenly around the other three sides, the base is placed face down on a flat surface such as a glass slab and the line is extended around all sides with a fixed pointer calibrated to the height of the line just incised. The base is turned upright and two concentric lines are traced out on its top, one 6 millimeters from the edge and one 4 millimeters from the edge (8). The entire area inside the inner line is carved out (9). A sharp blade is used to finish the inside; the walls and bottom of the base are left a little thicker than those of the lid (10). With a straightedge as a guide, the flange is carved out (11). The fit is checked and any necessary adjustments are made, possibly including an attractive beveling of the corners (12). The piece is dried slowly with the lid in place.

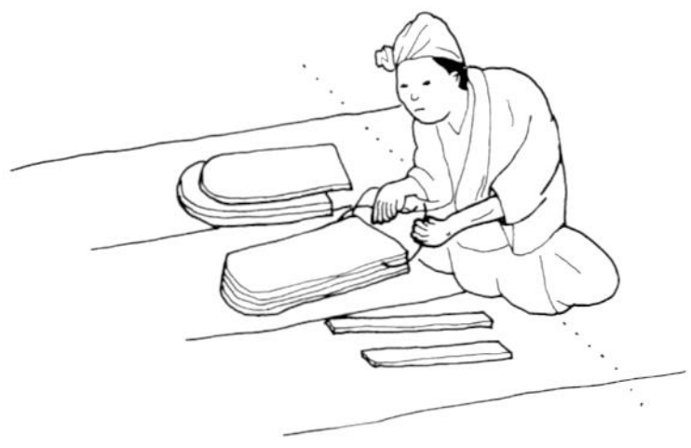

45. Cutting slabs

Slab Making

The traditional method of making slabs is to cut horizontal slices through a lump of clay with a wire or cord (fig. 45). The above-mentioned measuring sticks are used as a guide. Alternatively, a string is stretched between a pair of standing sticks with ruled notches, and the height of the string is adjusted after each cut. Kyoto potters have been particularly proficient in slab construction since the late seventeenth century, when they began to make ceremonial stands, stacked boxes, and imitation fans, objects traditionally made of thin slabs of wood or even paper. The local potters call the method "cabinet work."

Prior to cutting slabs, the clay is kneaded very well. Especially when making objects with flat surfaces, the clay should be kneaded from both ends of the lump.

Slab cutting (fig. 46). The clay is slapped with the hands or a paddle into a block approximating the intended shape (1). The block should be roughly twice as long and twice as wide as the finished piece, and its height should be sufficient to produce the necessary number of slabs. Corners are fashioned on the mass before cutting (2). The measuring sticks are then stacked along the long axis of the clay. The cutting wire is pulled tautly across the far end of the clay mass, with the thumbs pressing the wire firmly against the top of each stack of sticks (3). The wire is pulled through the clay (4). To keep the sticks from sliding, their ends are jammed against the thighs. After each cut, the top stick is removed from each side (5), and the next cut is made. Slabs from the very top and bottom of the mass generally are not used. After the required number of slabs are cut, the sticks are removed and the slabs are gently peeled off one by one (6). The slabs are laid on flat boards that have been lined on both sides with paper, and then they are stacked to dry (7). If the abrasions left by the cutting wire are objectionable, they are smoothed with a flat rib (8).

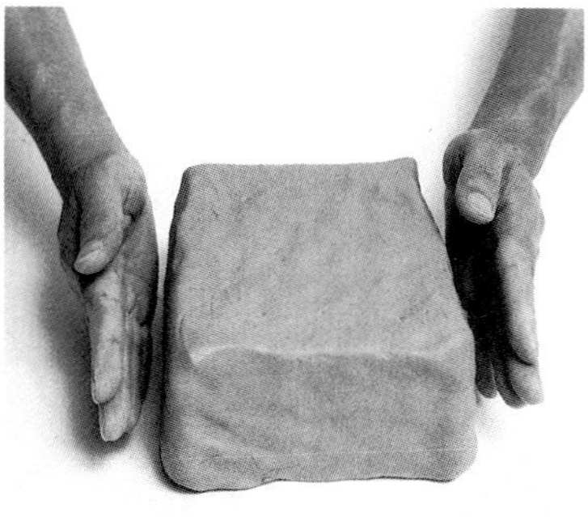

1

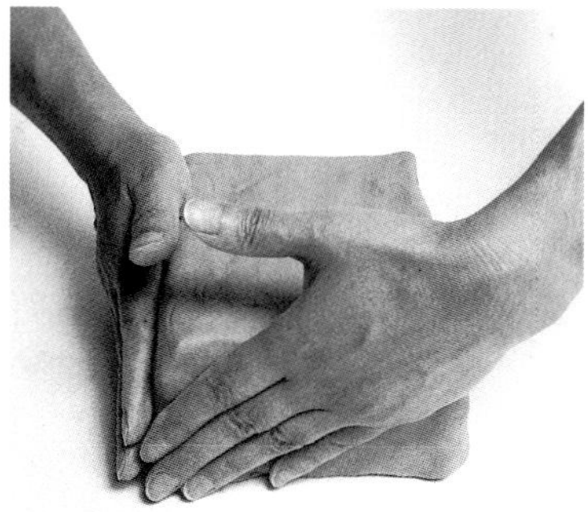

2

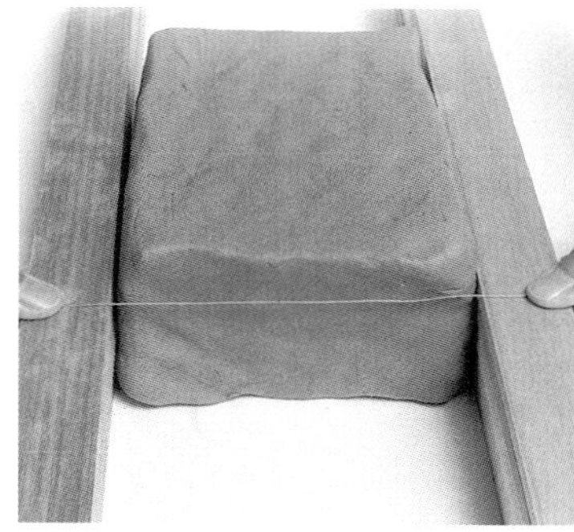

3

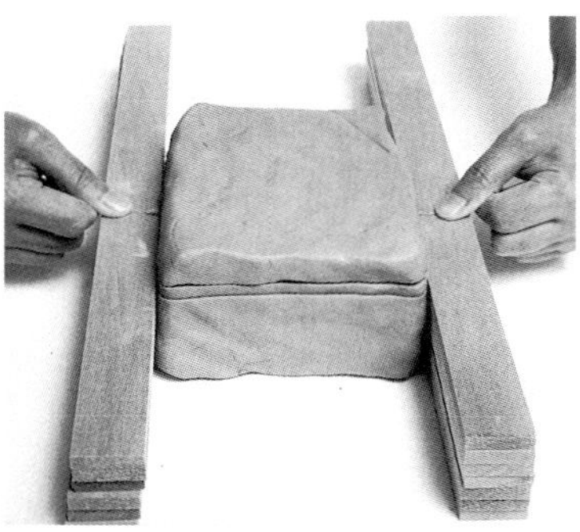

4

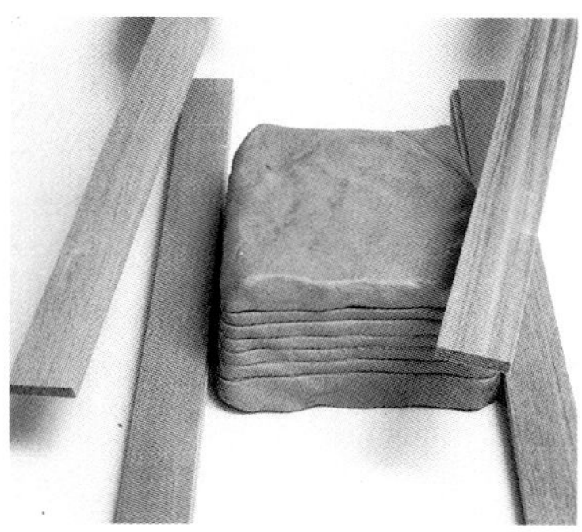

5

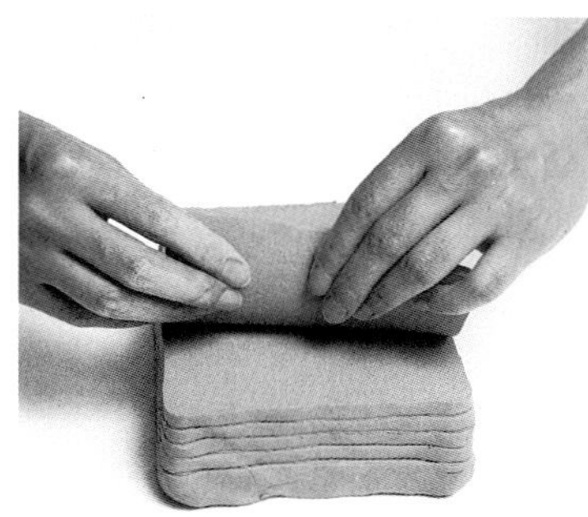

6

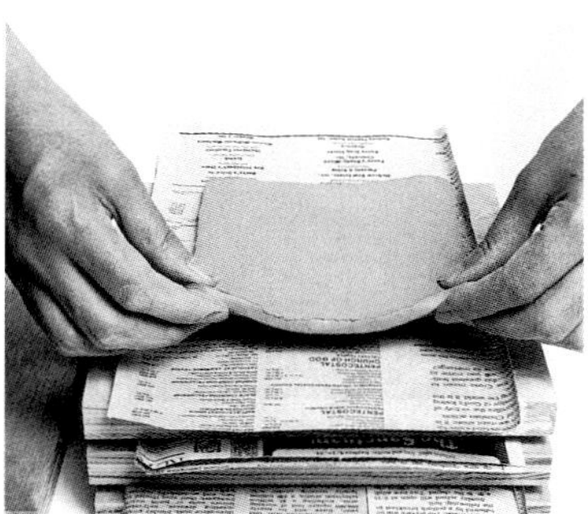

7

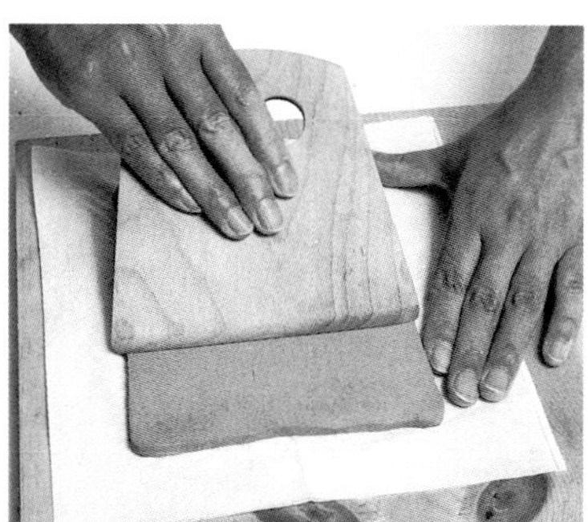

8

46. Slab cutting

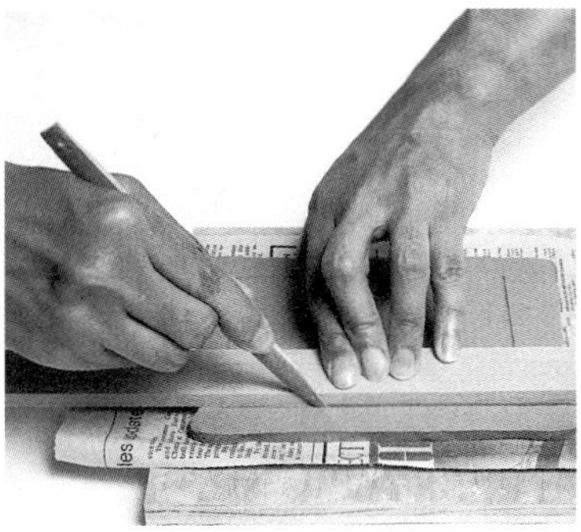

1

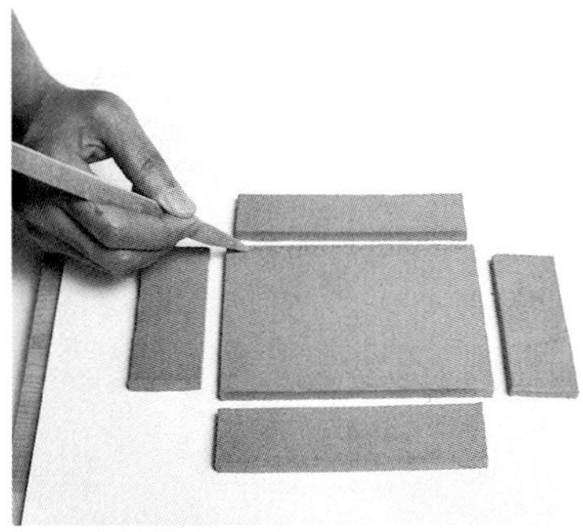

2

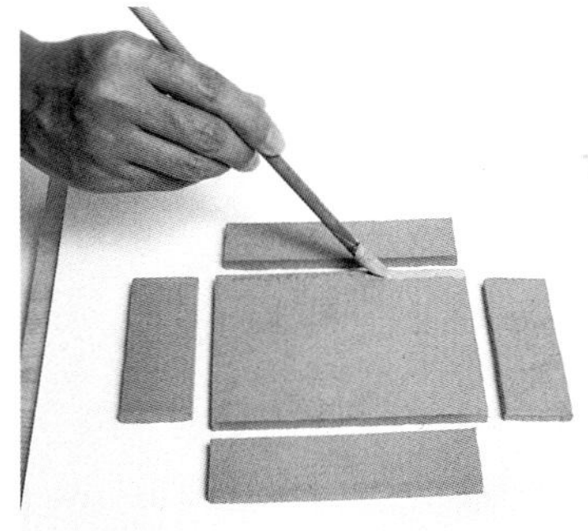

3

47. Slab-assembled box

Slab Assembly

For rectilinear objects, the slabs are dried by sandwiching them between flat boards that have been lined with paper, as in the above procedure. After the slabs have dried enough to hold their shape, the boards are removed.

Slab-assembled box (fig. 47). The base of the vessel is cut out using a ruler as a guide, to ensure that all cuts are perpendicular(1). The side walls of the plate are cut out. Since these walls will rest on the base, the thickness of the base will figure into the calculation of the height of the walls. The perimeter of the base is then scored with a needle or combing tool (2). A slip made out of the same clay body as the slabs is applied to the scored surface (3). A side wall is attached and its joint reinforced with a thin coil pressed into the internal seam (4). The wall is kept in place by supporting it with a stack of measuring sticks on the outside. The reinforcing coil is smoothed (5). The other side walls are added, and their corners are smoothed in the same manner (6). The exteriors are smoothed with a sharp, flat-edged rib (7). Using the measuring sticks as a guide, the rim is trimmed with a sharp knife (8). A crisply formed box is the result (9).

This procedure is simple enough in the raw clay stages, but the problem occurs in drying and firing, when the clay shrinks and warps. Three precautions will help avoid this problem. First, the original clay mass should be well kneaded and rigorously hand-

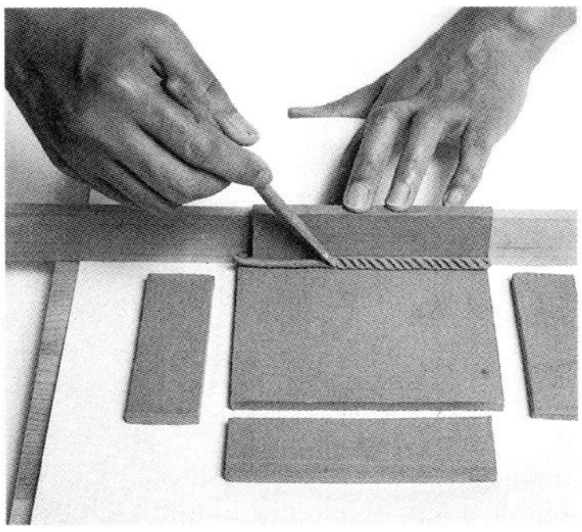
4

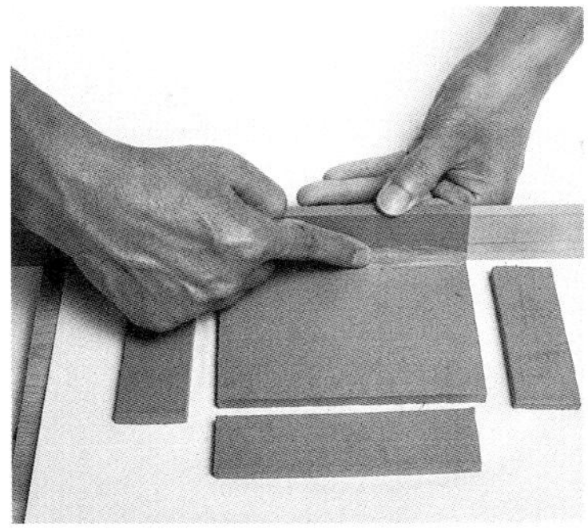
5

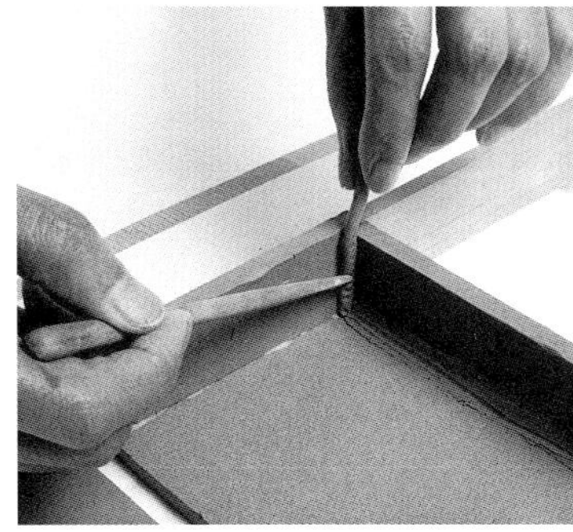
6

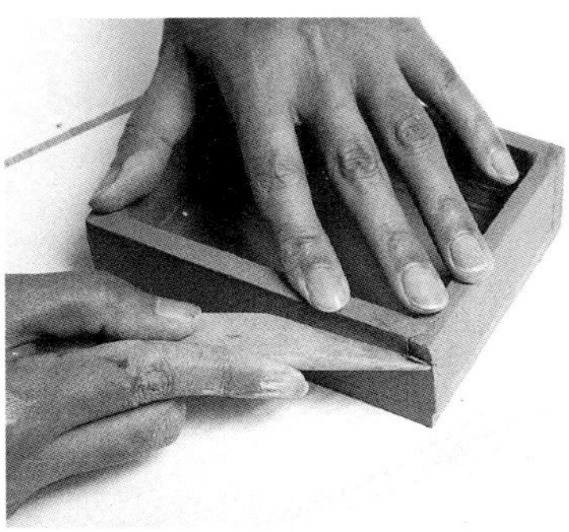
7

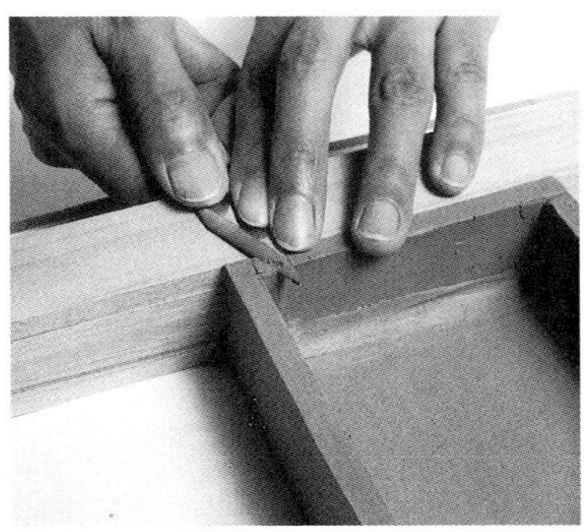
8

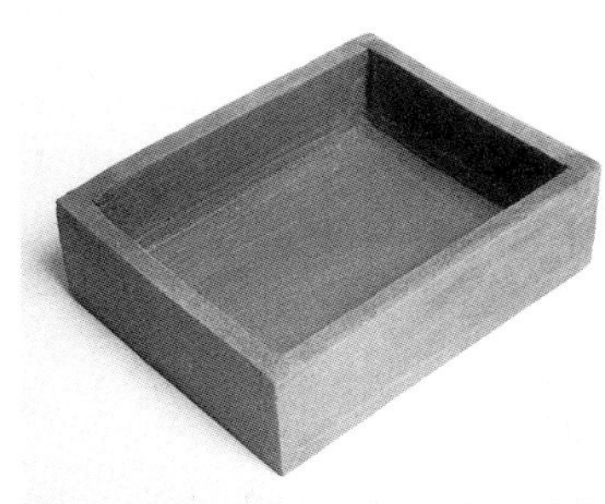
9

paddled prior to the slab cutting. Second, the side walls should be formed with an ever-so-slight outward bulge, which will later shrink into a straight edge. Third, the piece should be dried very slowly in a wet-box or wet-room and taken out and gently paddled if there is any warping. Large boxes, even porcelain ones, can be thickly modeled and then trimmed gradually over a long drying period. In this way irregularities are eliminated as they appear.

If a lot of slab-assembled pieces of the same size are made, bats are cut out to match individual parts. These are used as templates for cutting out the slabs and provide a convenient external reinforcement when the pots are being assembled. Such templates are used in the Aizu-Hongo pottery for the assembly of their well-known pickling boxes.

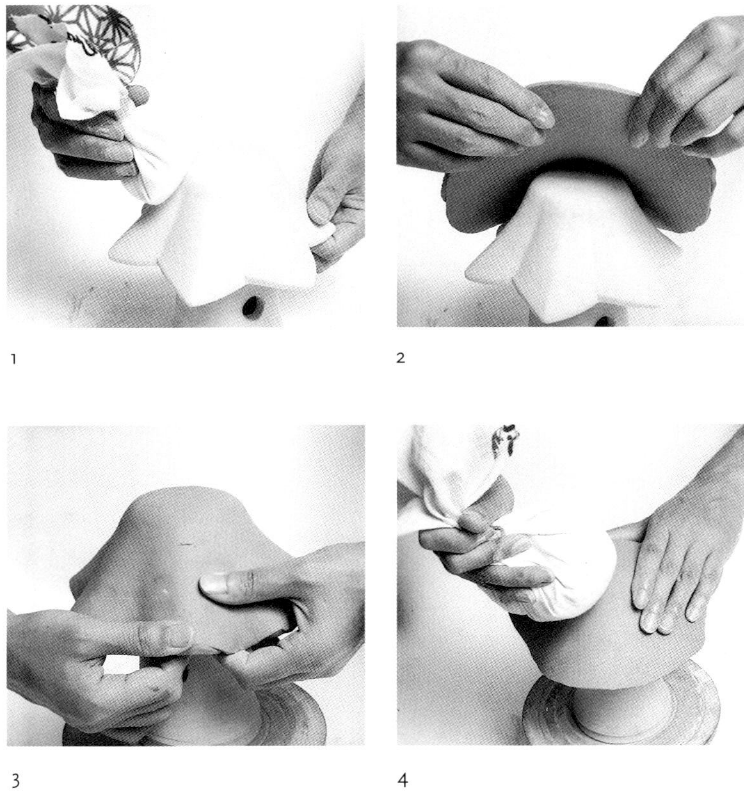

48. Drape-molded bowl

Drape Molding

Today molds are generally made of plaster, but the traditional materials were wood and bisque-fired clay. Drape molding achieved special prominence in the diverting, idiosyncratic Oribe-ware food vessels of the early seventeenth century (see pl. 10), and the irregular shapes made possible with the drape-molding process continue to be much admired in Japan.

Drape-molded bowl (fig. 48). The mold (here for a flower-shaped bowl) is placed on a banding wheel or turntable and the outside is powdered with a starch-filled dust bag (1). A slab is cut and draped over the mold (2). The clay is gently pressed down from the center of the slab working outward. All gaps between slab and mold are eliminated. The edges are the most difficult place to form: they are

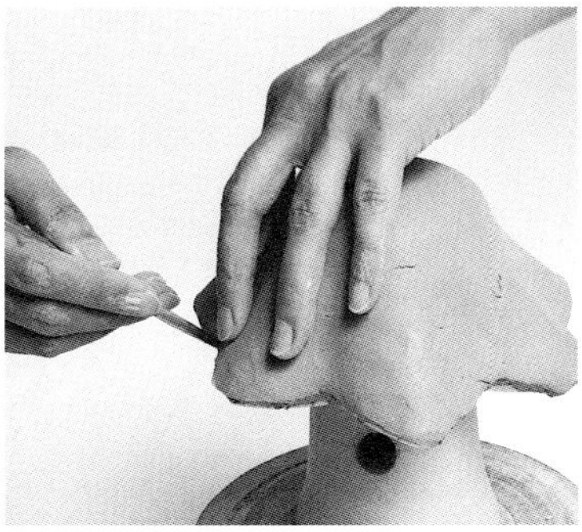

5

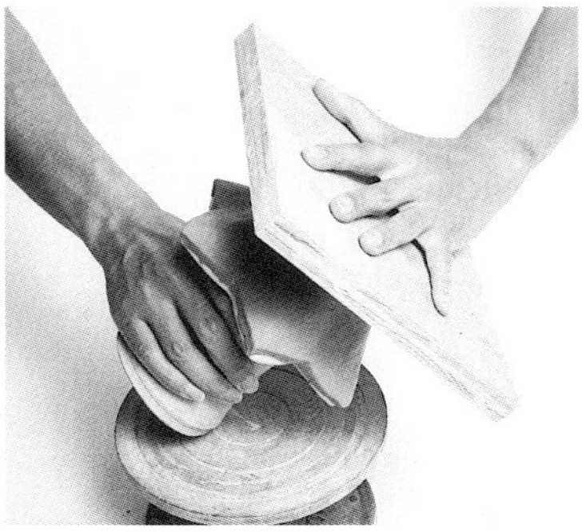

6

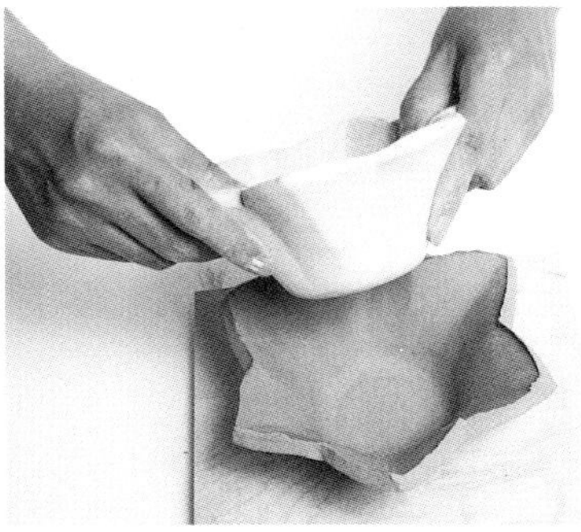

7

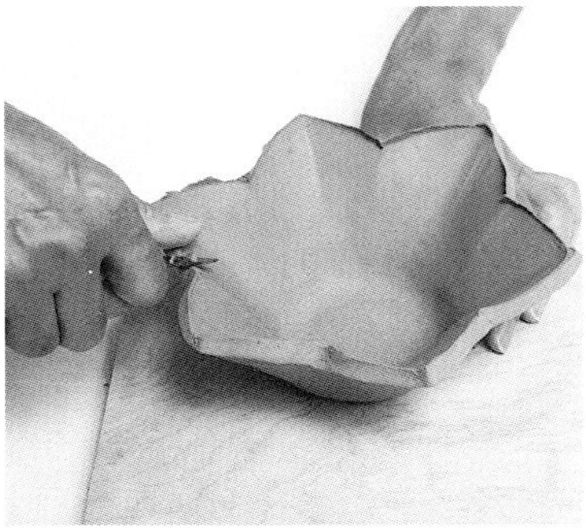

8

pressed firmly against the mold with downward strokes of the thumb, which shears off most of the excess clay (3). For final shaping, slab and mold are covered with a soft cloth. Care is taken that the cloth doesn't wrinkle, which will mar the surface. The slab is pressed gently but firmly against the mold with a grog- or sand-filled bag, again working from the center outward (4). The cloth is removed and the edge of the pot is trimmed with a knife or trimming bow using the edge of the mold as a guide (5). Any irregularities are filled in with clay. The form is dried for 10 to 20 minutes. A flat board is placed on the base, pot and board are turned over (6), and set right side up. The mold is lifted out (7). If the mold sticks, it is set aside until it comes out easily. The rim is cleaned up with decisive cuts from a sharp knife (8) and then finished with a moistened chamois.

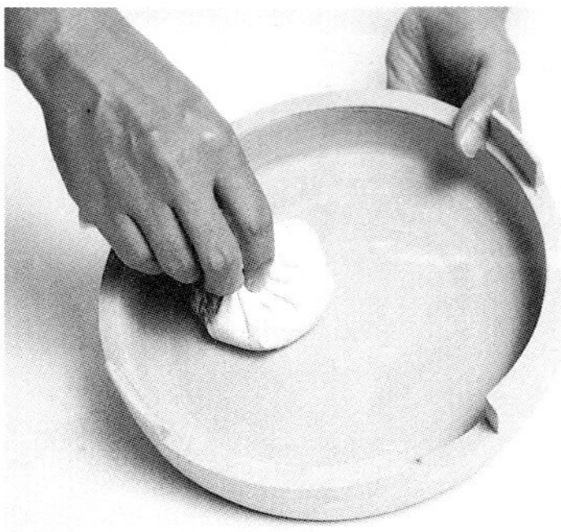

1

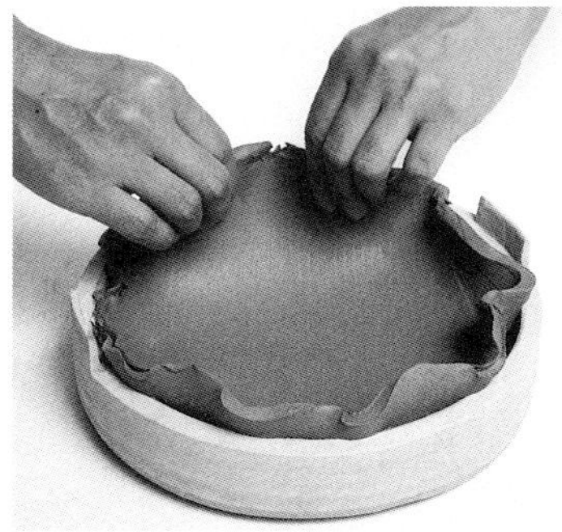

2

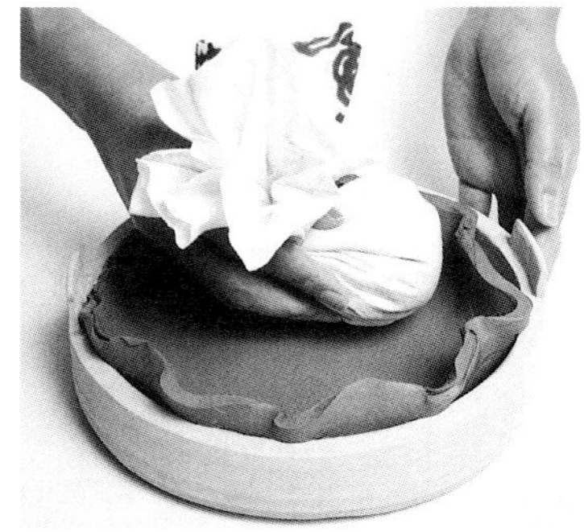

3

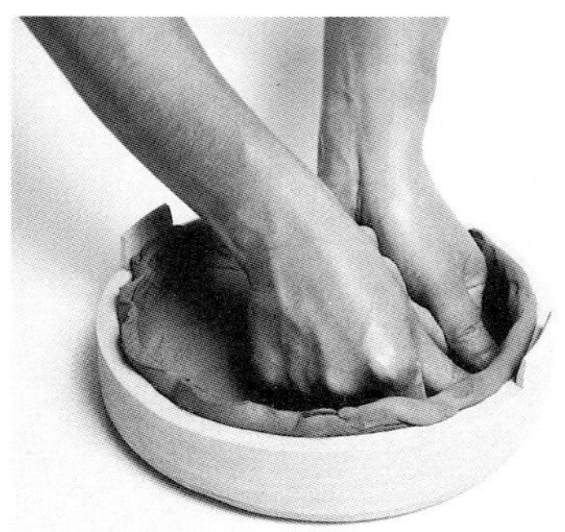

4

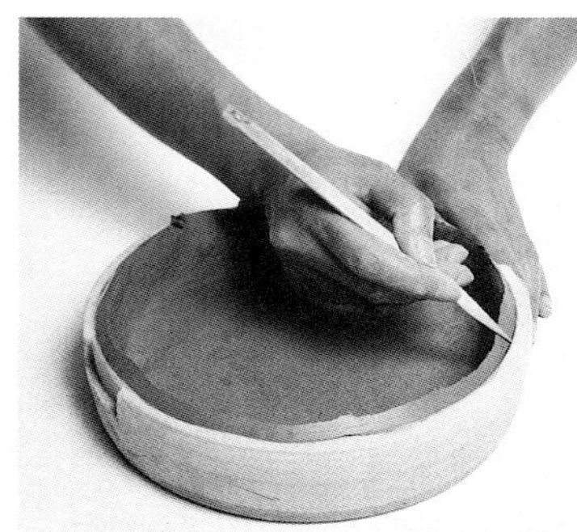

5

49. Piece-molded vase

Press Molding and Piece Molding

These techniques are used to make the press-molded bottles made famous outside of Japan by Hamada Shōji. The method requires a two-piece, keyed mold.

Piece-molded vase (fig. 49). The inside of one mold is dusted with the starch bag (1), and a slab is draped into the mold (2). A bag filled with sand or grog is used to tap the clay into the mold (3), and the clay is pressed firmly into the corners with the fingers (4). The excess clay is cut away, but a little ridge is left protruding above the edge of the mold (5). These same steps are performed for the other half of the mold. When both slabs are in place, the protruding edges are moistened (6) and the molds are joined (7). Sometimes they are bound together with a heavy rubber band or piece of inner tube. The inner seam is smoothed with the fingers or with the throwing stick (8). The mold is stood right side up, and after 10 to 20 minutes, the mold

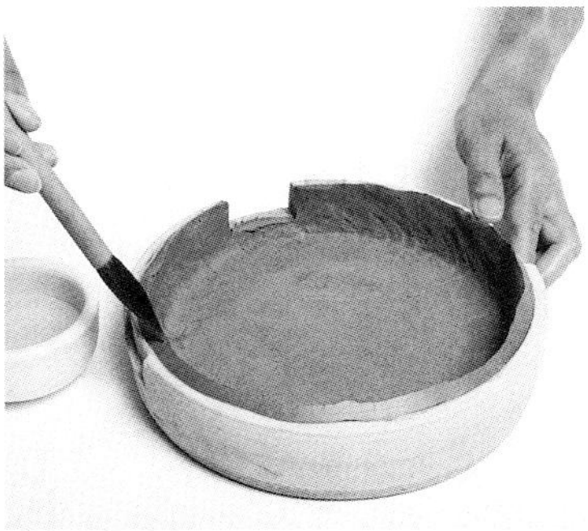

6

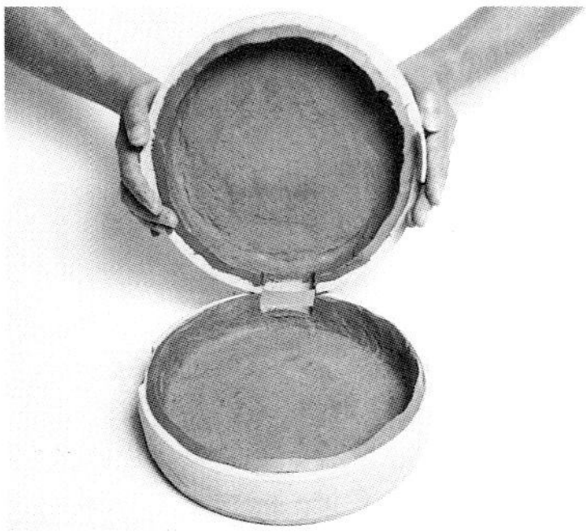

7

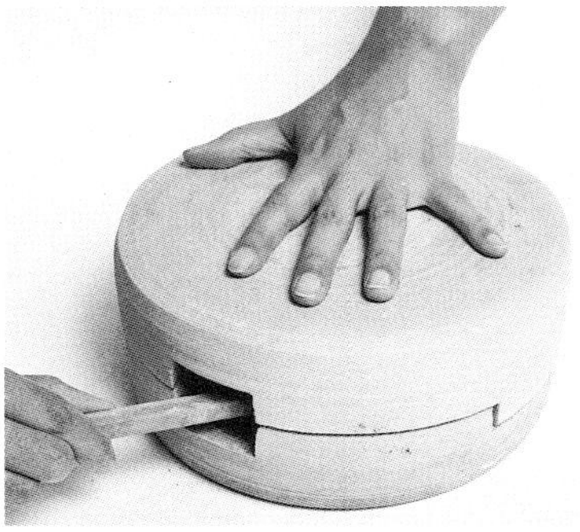

8

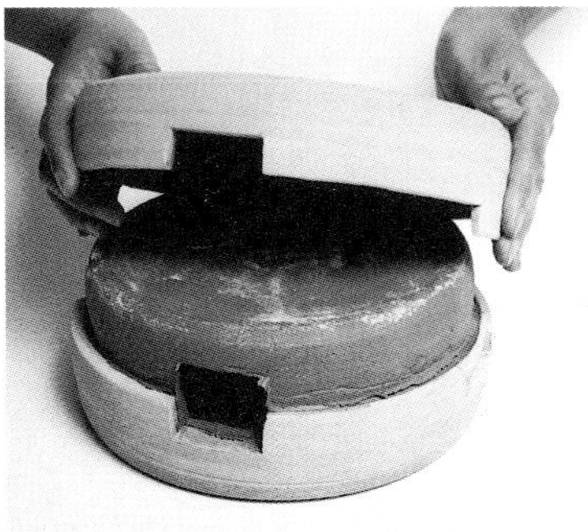

9

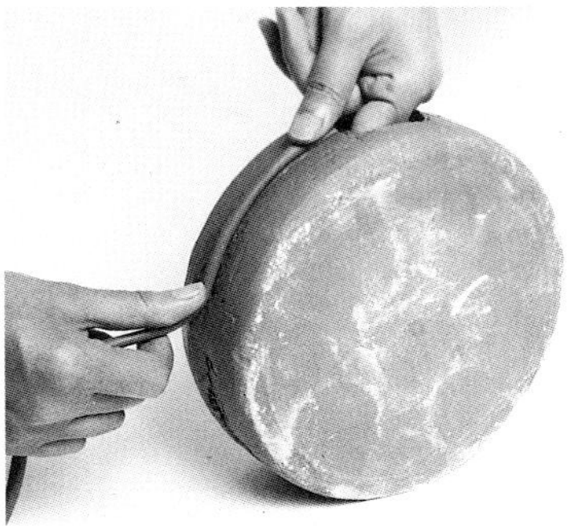

10

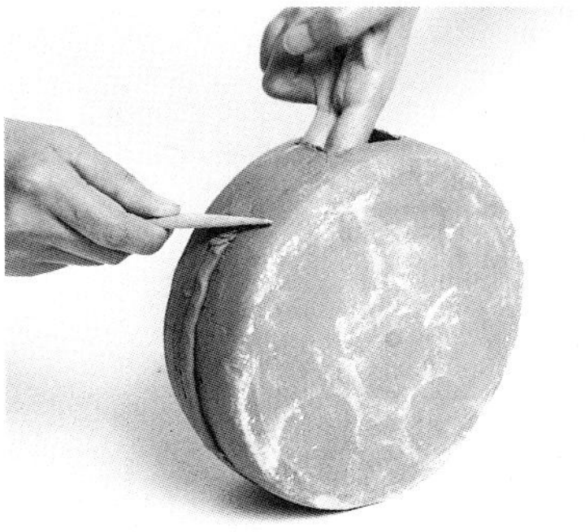

11

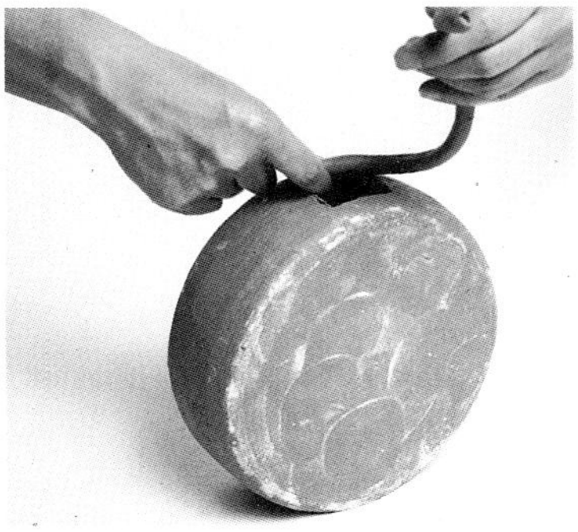

12

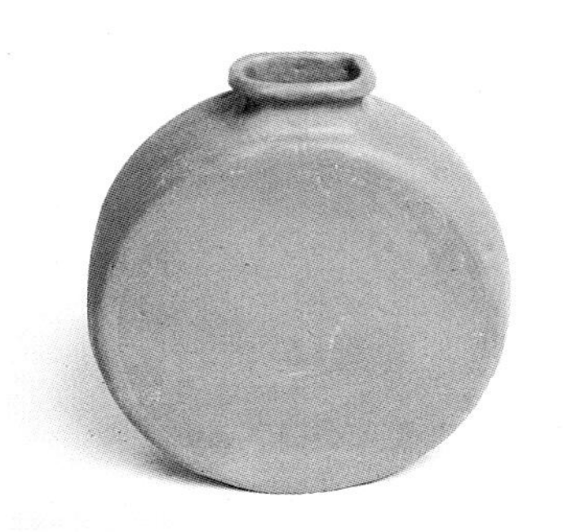

13

halves are pulled apart (9). To smooth the external seam, a thin coil is applied over it (10), and the area is trimmed with a blade (11). A mouth is fabricated for the vase (12). The bottom may be paddled to create a flat base (13), or a foot may be added.

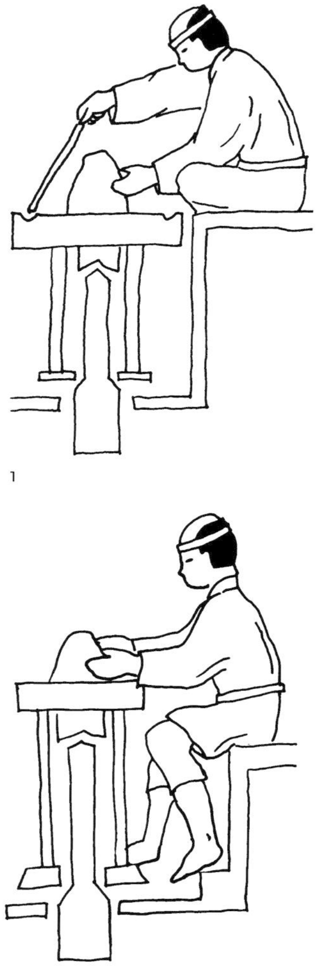

1

2

50. Hand wheel and kick wheel

WHEEL TECHNIQUES

Excavations have revealed that prehistoric potters revolved their pots on mats and leaves prior to the introduction of the potter's wheel around the fifth century of the common era. The fast-turning wheel offered great advantages, especially for uniformity and speed in construction. The traditional wheel types are the hand wheel and kick wheel (fig. 50). The hand wheel (1), propelled by a stick held in the right hand, is a sensitive tool for smaller wares, where one hand can do much of the forming. Kyoto potters, who used the hand wheel until recently, even took pride in never dirtying their right (turning) hands. The kick wheel (2) leaves both hands free and thus is advantageous for throwing larger forms. For very large pieces, artisans work in teams: an assistant turns the wheel by hand, by a pulley-like system (fig. 51), or by kicking (fig. 52). Electric wheels are now the overwhelming favorite, but one can still find potters turning their own wheels with astounding skill. Some potters believe that the slow, irregular turn of these wheels gives the pots an organic feel, but more likely it is the degree of control that is most appealing. In most cases the wheel turns clockwise, although some potters in Kyushu trim with the wheel turning counterclockwise.

Even though there are a good number of books that teach the use of the wheel, it is not easily explained in words and pictures. One has to work through the procedures repeatedly. Nevertheless, most people can grasp the basics after four or five days of work. Training proceeds from first learning how to center a large mass of clay by raising and lowering it and then by making small things like teacups and dishes "off the hump," as the practice is called. After

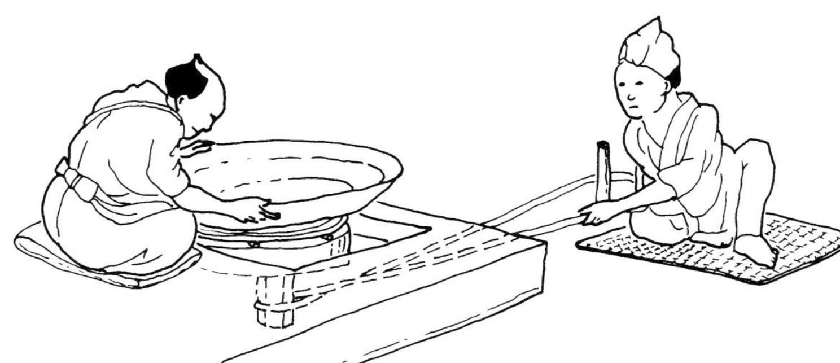

51. Wheel with "pulley turning" by assistant

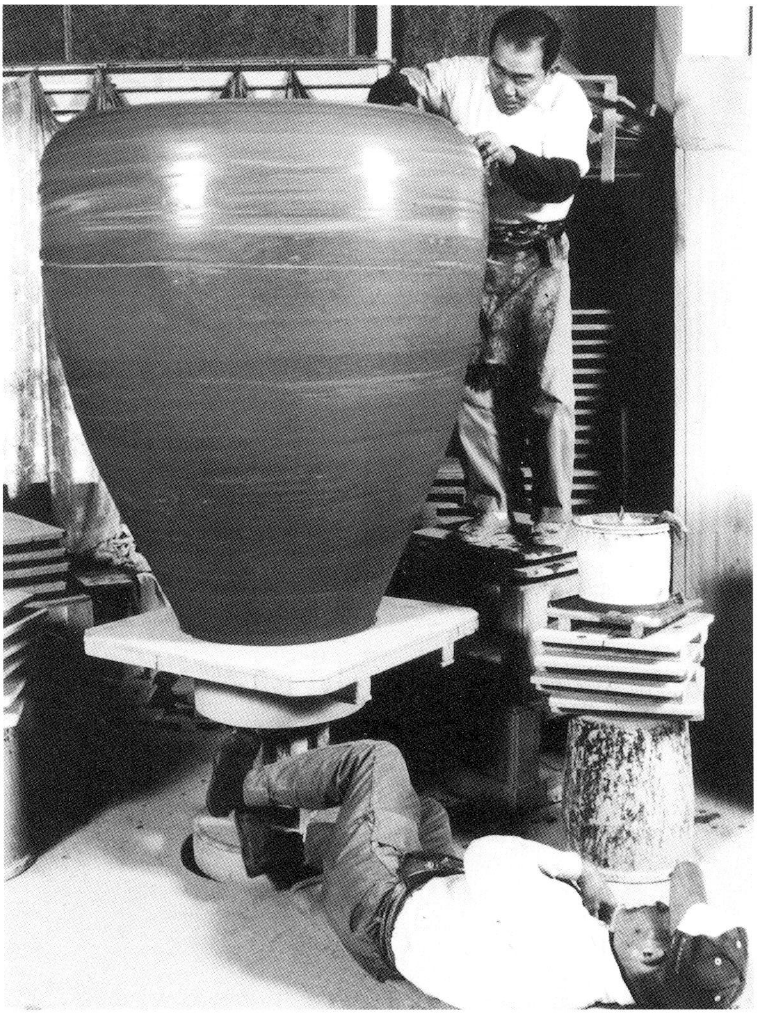

52. Wheel kicked by an assistant

obtaining proficiency in making small objects off the hump, the student or apprentice moves on to larger objects like dishes and jars.

Since throwing off the hump differs from the traditional Western manner of working from small balls of clay that have been sectioned out prior to throwing, and since some details of Japanese throwing are different from Western ones, the basic procedures are covered in the following sections.

Centering

Hand-kneading, as described above, drives out all the air pockets and evenly distributes moisture and density. But when forming on the wheel, even the slightest irregularity left over from the kneading will affect the throwing. If the potter raises and lowers the clay mass under pressure, the clay becomes centered on the wheel and

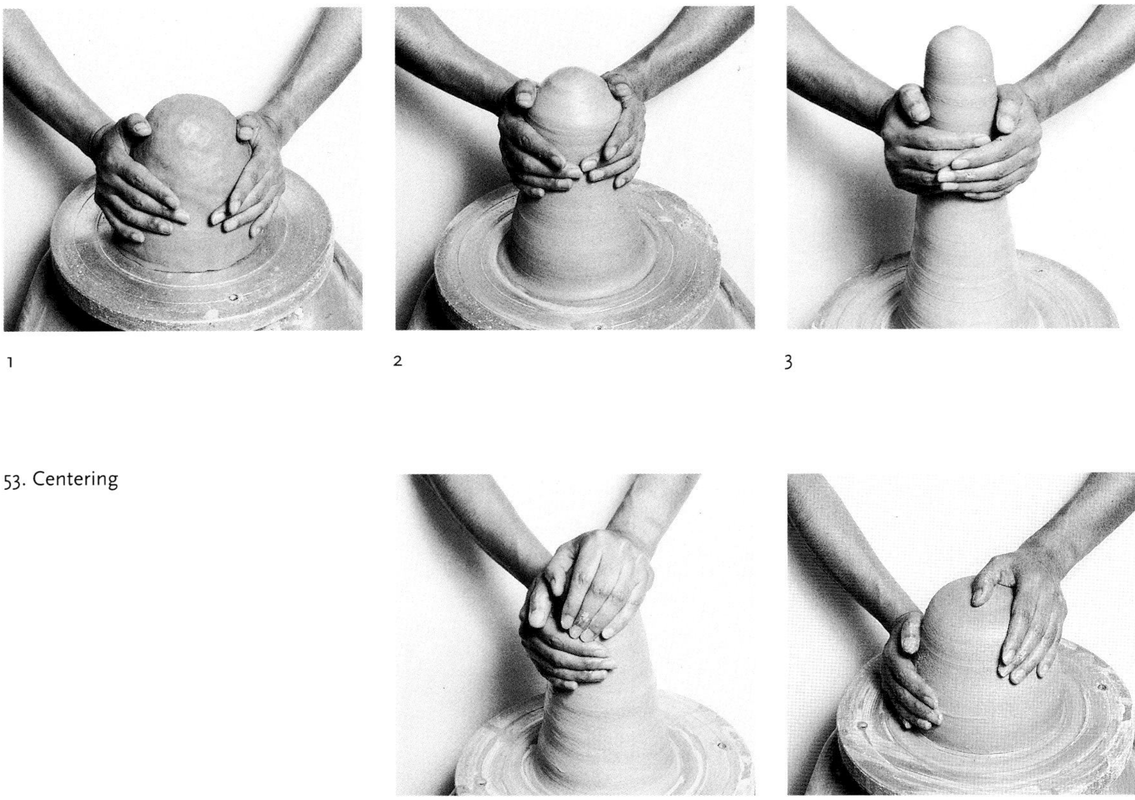

53. Centering

thoroughly homogeneous in consistency. Centering is also necessary to achieve walls of even thickness and height. The procedure calls for an intelligent application of a small amount of strength, rather than brute force.

Centering (fig. 53). A ball of well-kneaded clay is placed on a moistened (not wet) wheelhead. With the wheel turning slowly, the ball is slapped down onto the wheelhead with a clapping motion (1). The hands are soaked in water (many potters use slip) and applied to the clay. The wheel speed is increased. With elbows braced on thighs, the potter compresses the clay inward from both sides. In response, the clay mass thins and rises, and the hands rise with it, maintaining firm, even pressure (2). As the mass tapers to a point (3), pressure is reduced and the hands are withdrawn. The entire clay mass is affected by this thinning. If only the surface of the clay is pulled upward, a soft well will form in the center of the clay mass. If that happens, the top of the mass must be cut off.

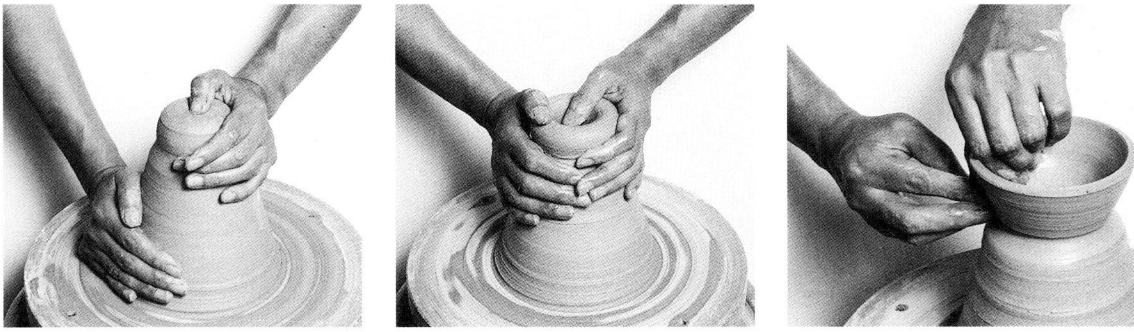

1 2 3

54. Small cup

To work the clay back down into a compact, centered mass, the left hand is placed on top of the clay and the right hand is set against the side of the mass. The left hand pushes down and away on the top, while the right recenters at a lower point on the mass (4). Steady and even pressure is necessary or the clay will go off center. If there is surface pressure only, a bulb will form on the top. The downward push is continued until the mass is back at its original height (5). This raising and lowering is repeated about four to five times, at which point the clay is ready for throwing.

Throwing Small Pots

The beginner usually starts with a small cup. This teaches the mechanics of throwing and provides a feel for how much clay should be sectioned off the hump to make a given pot. As mentioned above, small forms can be made with only one hand.

Small cup (fig. 54). The clay is raised and lowered as described above, and then the entire mass is tapered so that its top is about 5 to 7 centimeters across. The requisite amount of clay for the pot is sectioned off by grabbing the top of the spinning clay with the left hand and pinching inward with the middle and fourth fingers to make a knob (1). This is the amount that will go into the vessel. Skillful sectioning-off is essential to the manufacture of identical pieces. To begin the throwing itself, the top of the clay mass is flattened with the thumb of the left hand and then the thumb is pressed down to form a well. A wall forms between the thumb and other fingers (2). At this point, the wall can be pulled up with one hand. The hand is shifted to a vertical position with the thumb going into the well and the other fingers along the outside wall, and the wall is

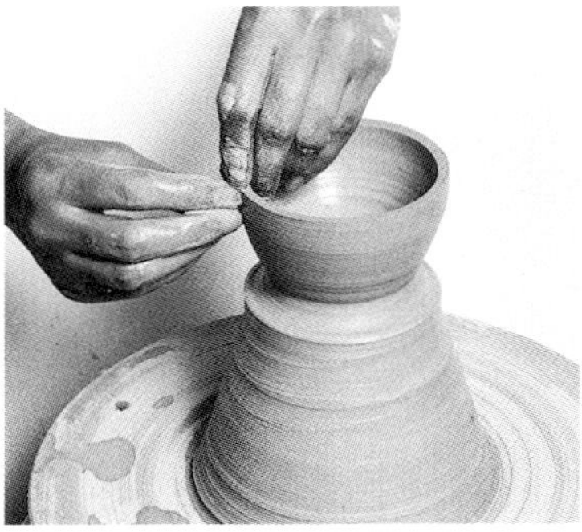

4

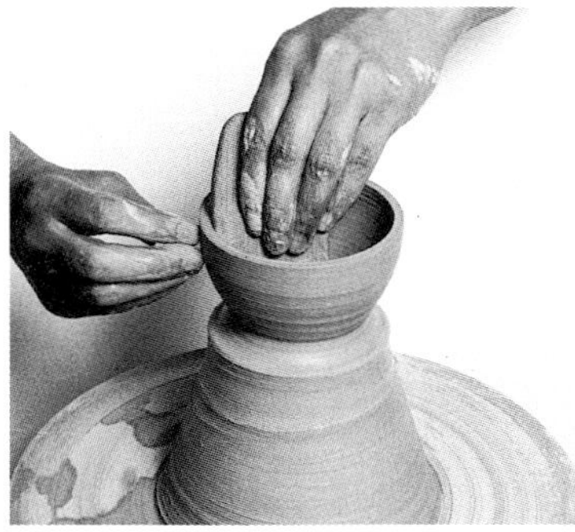

5

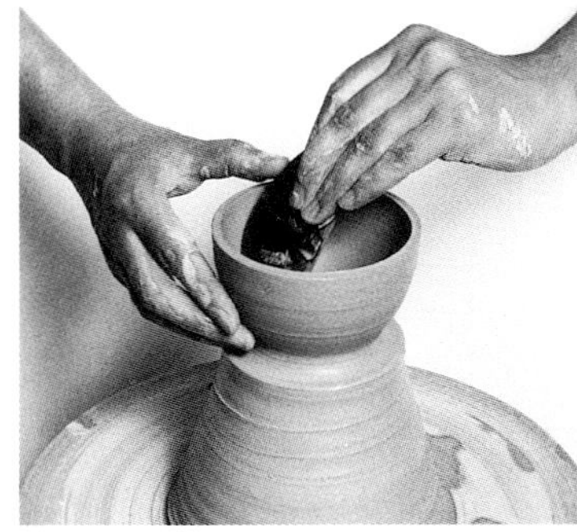

6

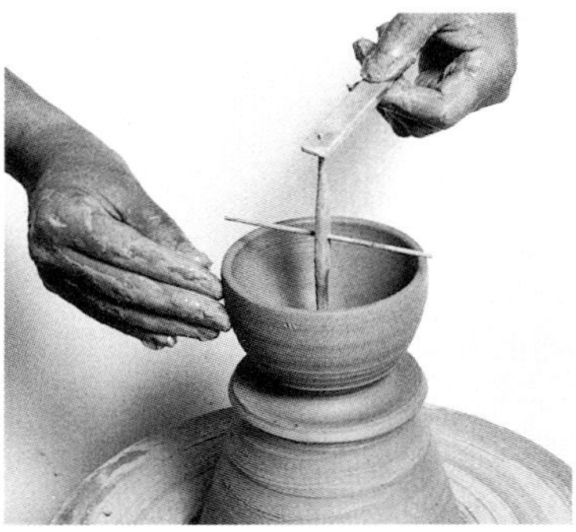

7

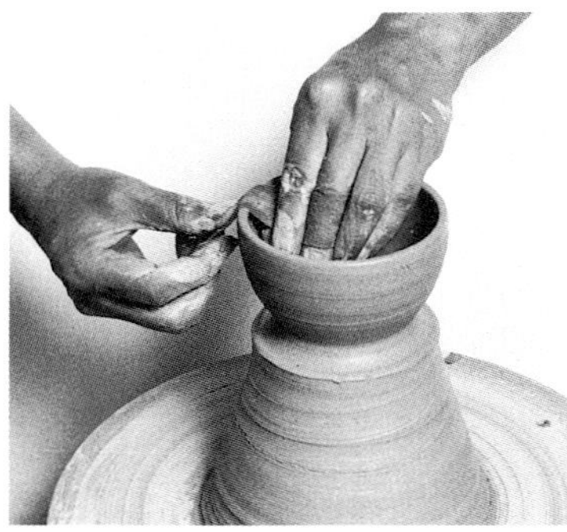

8

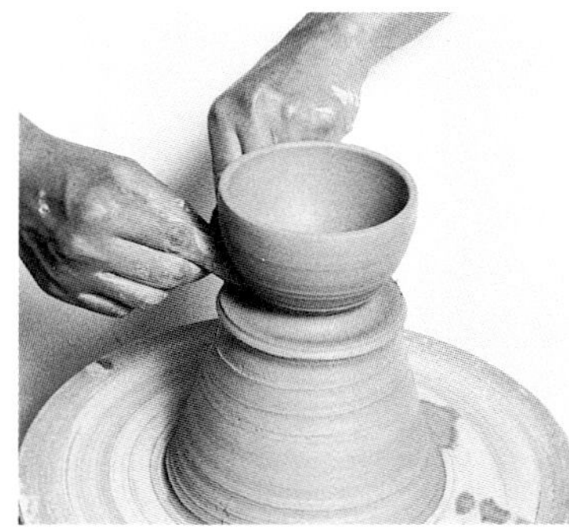

9

54. Small cup (continued)

pulled upward in one decisive motion. This is the one-handed throwing procedure mentioned above.

For two-handed throwing, which is necessary for larger or more complicated forms, the left thumb is placed outside the wall, opposite the other left fingers on the inside, forming a "C" shape. The fingertips of the right hand support the exterior wall. The wall is pinched with the left fingers, which causes it to rise. The left fingers pull upward along with the rise (3). At the end of the pull, the left index finger moves to the top of the rim, consolidating it (4). In subsequent pulls, as the vessel becomes too deep for the left thumb to be outside, the left fingers on the inside line up against the right fingers on the outside and squeeze the clay between them. Both hands guide the subsequent rise. Pressure from the inside determines the basic contours of the vessel.

When the shape has more or less achieved its intended size, a throwing rib is used to finish the shape. For larger forms, a round-edged rib is used to approximate the shape, and a sharp-edged rib is used to finish the work. The rib is held in the left hand and placed

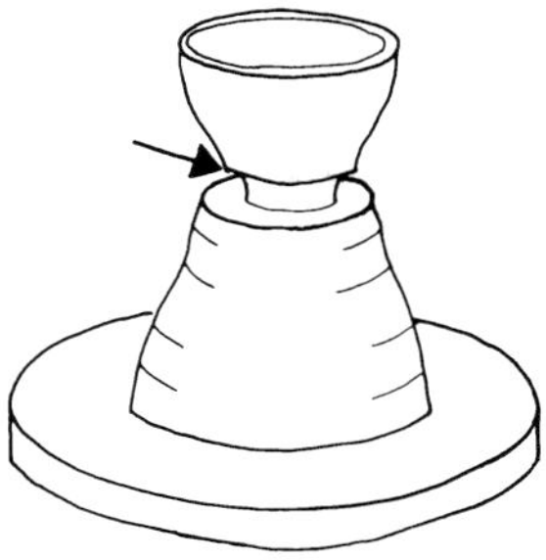

10

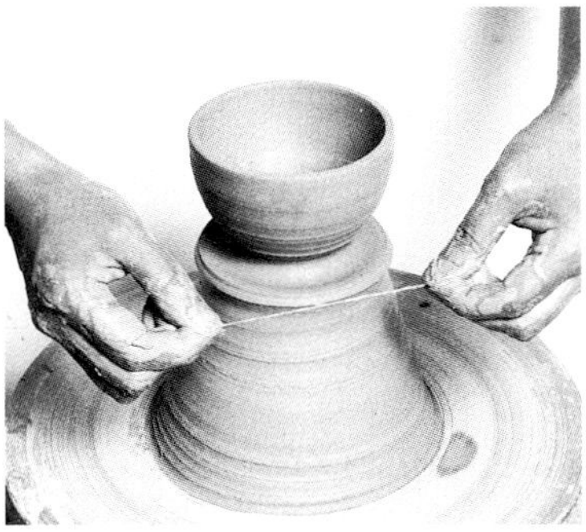

11

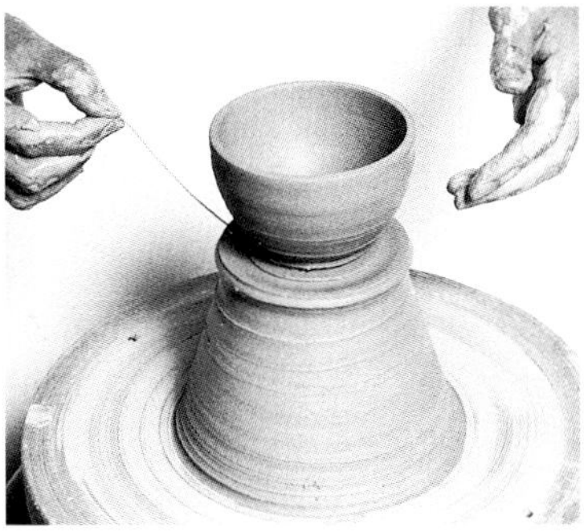

12

in the cup to the right of the central axis. Gentle pressure is applied outward and downward (5). This pressure is counterbalanced by the fingers of the right hand placed along the outside wall. The rib is usually tilted a little bit rather than held in a strictly vertical position (6), which keeps the blade from snagging the spinning clay. There should be no water, slip, or finger marks in the interior after the finishing rib is withdrawn. When the desired shape is achieved, the depth and diameter are checked simultaneously with the dragonfly gauge (7). Care is taken not to nick the bottom of the cup with the gauge. Necessary adjustments are made. For example if the cup is too deep, the rim can be sliced off with a needle or sharp spatula. The lip is then finished with a wet chamois. A loop is formed with the chamois between the index finger and thumb of the left hand and applied gently (8).

Prior to removing the pot, the clay below the base of the form is narrowed by pressing in slowly with a wet index finger (9). In effect, this creates a kind of stem for the piece (10), and the narrower the stem, the less distortion when the pot is removed. To cut the piece off the remaining clay mass, the cutting cord is held on the far side of the pot and drawn in until is it just lightly touching the turning stem (11). The left end of the cord is released. After it wraps around the stem once, it is pulled out with a quick horizontal jerk (12). The pot is severed from the mass in a second. (It takes some practice to find the proper speed of the wheel and the correct hand motion.) The pot is lifted from the base with the index and middle fingers of both hands and set on a ware board.

Cylindrical cup (fig. 55). These are especially difficult to throw. After making the basic form, the rib is firmly applied downward to establish the bottom of the cup as outer reinforcement is provided with

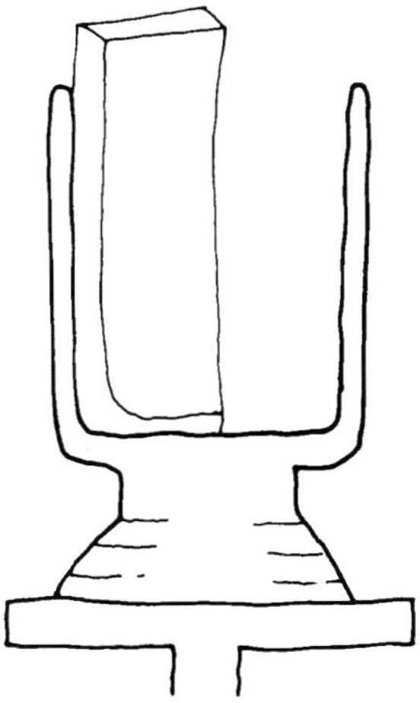

55. Cylindrical cup

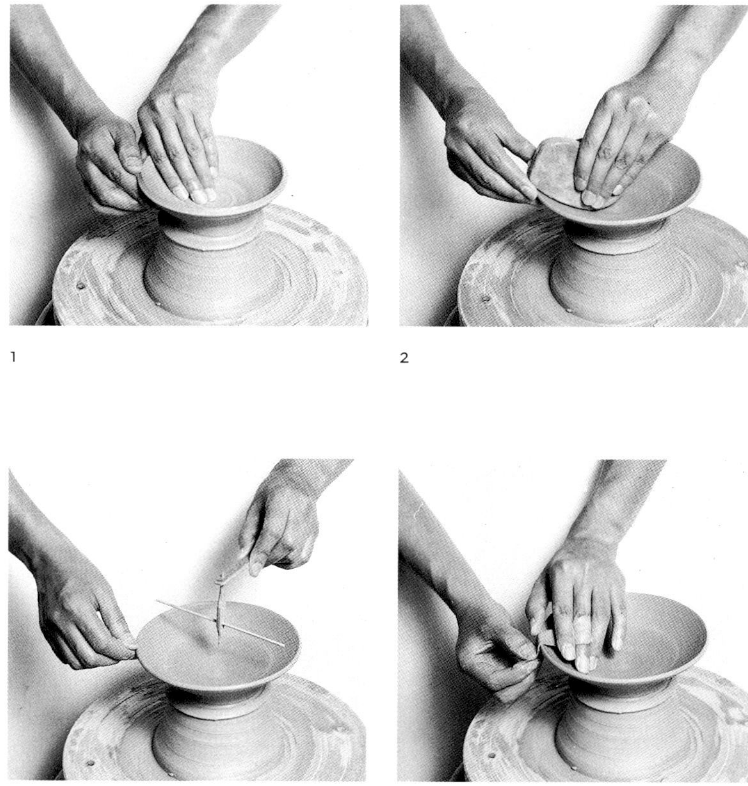

1 2 3 4

56. Small plate

the right fingers. The rib is pulled outward to form the corner. It is necessary to lift slightly when pulling outward, or the corners will become deeper than the center of the cup. The walls are pulled upward, with the rib inside and the right fingers on the outside. Excess slip is removed with the finishing rib. The cup should appear just a little narrower at the bottom than at the rim. If a slight bulge is left just below the rim, the sides will fire straight.

Small plate or dish (fig. 56). Flatware is slightly more difficult than cups, although the procedure is basically the same. The same introductory steps are followed as for the teacup, but the walls are pulled in a more horizontal direction (1). Once the edges of the plate are flared out, there is no going back, because the edges tend to collapse from being overworked or from the application of too much water. The other problem is a center that is too deep. These problems are avoided by using the forming rib in the beginning stages (2). When the forming is finished, the diameter is checked with the dragonfly gauge (3) and the rim is finished with a chamois (4). Plates warp eas-

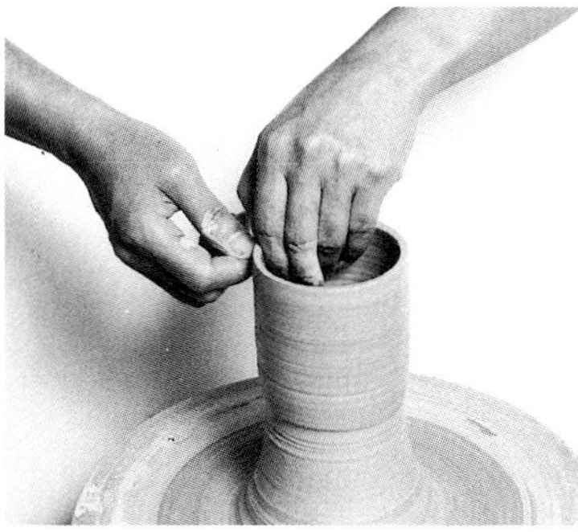

1

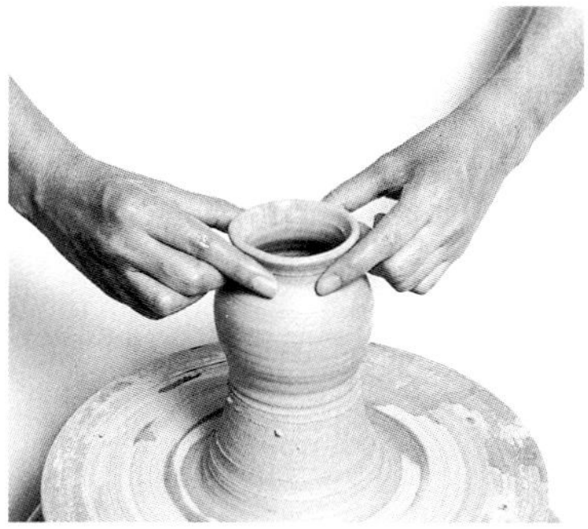

2

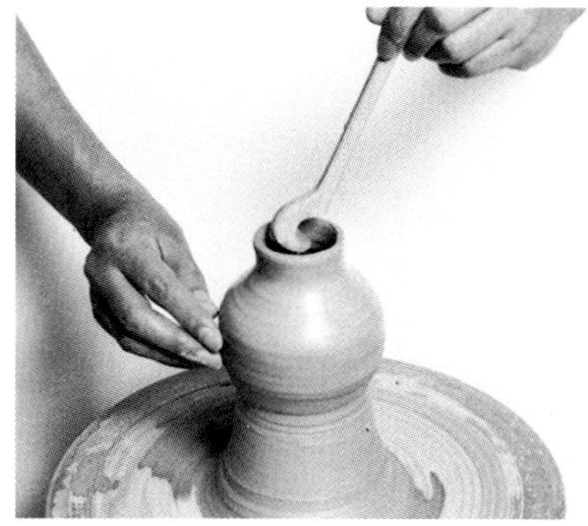

3

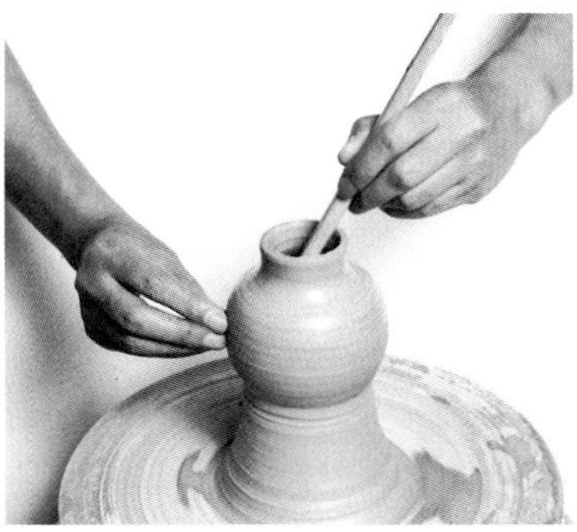

4

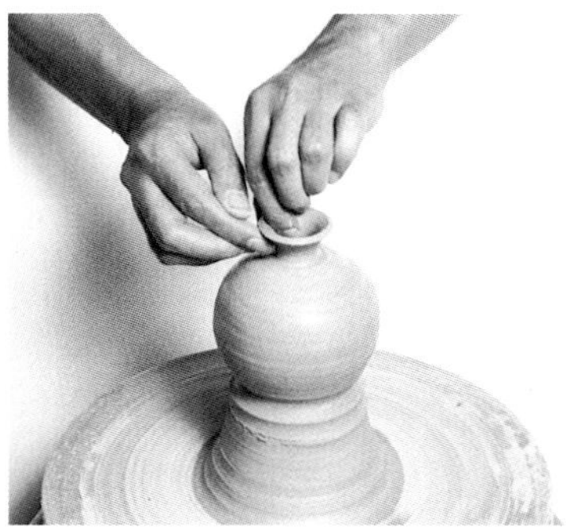

5

57. Small bottle

ily when they are taken off the wheel, but this can be avoided by pinching a narrower or longer stem below the plate prior to cutting.

Small bottle or vase (fig. 57). A knob of clay is sectioned off the spinning clay mass in the same way as described above. The amount should be somewhat greater than that for a cup of the same size. The left thumb sinks a well into the top of the clay. The left hand is positioned inside and the right hand outside, and the clay is pulled up into a cylinder (1); two or three pulls may be necessary. A neck is formed by pinching with both hands (2). This is done gradually to keep the clay from buckling. Just enough space is left for the throwing stick, which is inserted into the bottle. The stick first compresses and finishes the bottom contour and then is moved up along the wall. The course of pressure determines the profile, while the right hand reinforces the outside during this step. The throwing stick is removed when the basic shape has been achieved, and the neck and rim are straightened (3, 4). The neck is given a final narrowing with the fingertips—slip rather than water is the preferred lubricant dur-

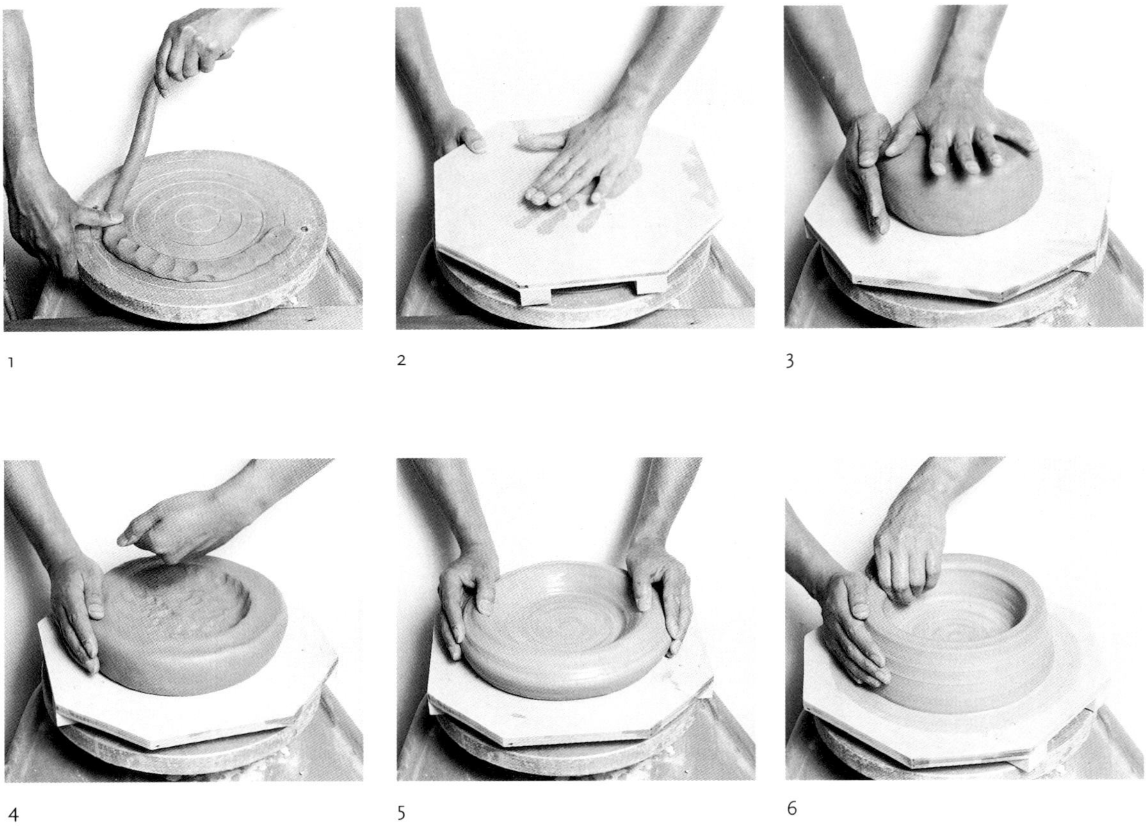

58. Preparing to throw large pieces

ing this stage—and then the lip is finished (5). The pot is cut off in the same way as the cup. When making a series of identical pieces, the girth of the bottles is measured with a wire caliper, using the broadest part of the bottle as the standard.

Throwing Large Pots

Large objects are usually made on bats attached to the head of the wheel.

Preparing to throw large pieces (fig. 58). The wheelhead and the cleats are first lightly moistened. The bat is secured to the wheelhead using coils of extra-firm clay in one of three ways: along the cleats, perpendicular to the cleats, or in a ring around the wheelhead (1). After the coils are applied, the bat is centered on the wheelhead and secured by pounding down on it with the palm (2). The top of the bat is lightly moistened, clay is placed on it, and the wheel is started slowly. The clay is slapped roughly into the center with both hands (3). Gradually it is patted down into a flat disc and then hol-

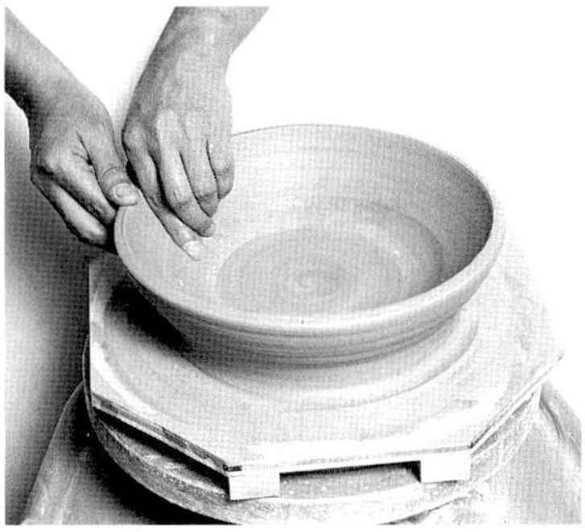

1

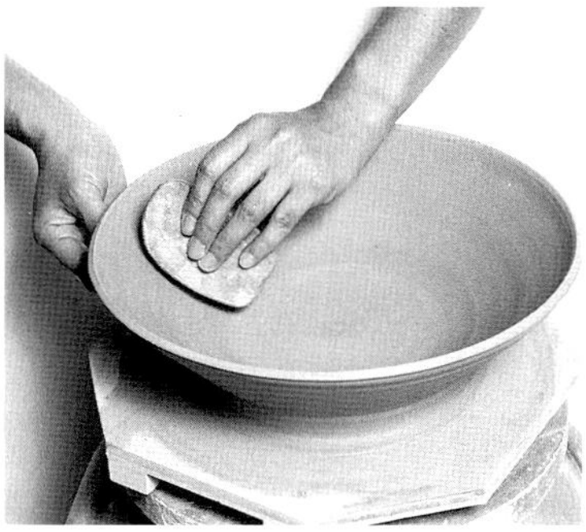

2

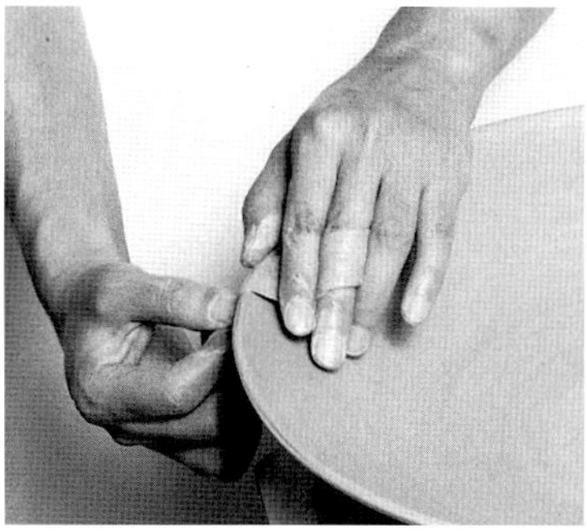

3

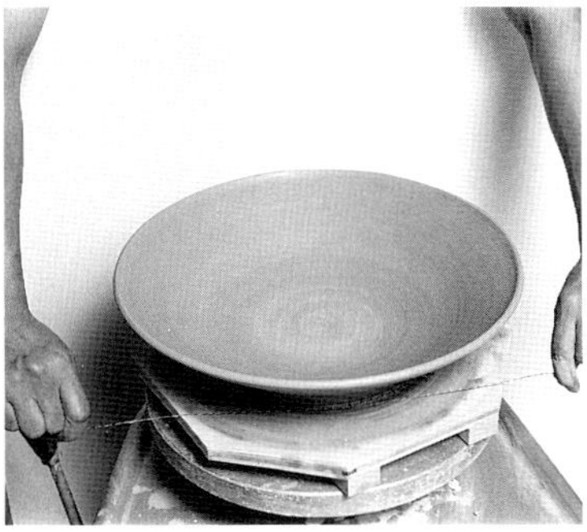

4

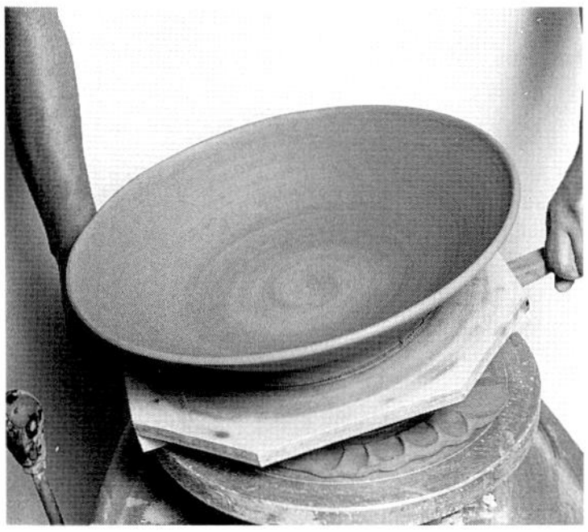

5

59. Large plate

lowed out in the middle with blows from the heel of the hand. The clay should spread somewhat beyond the intended base size of the vessel. The hollow is expanded by punching it with the fist (4), which also serves to compress the bottom. At this stage the thickness of the bottom can be checked with a needle. (The hole will disappear in the course of finishing the pot.) A thickness of about 3 centimeters is satisfactory. Water or slip is applied to the entire clay mass, the wheel is speeded up to about 60 revolutions per minute, and the mass is centered with both hands (5). With the right hand providing outer reinforcement, the left hand pulls up a low wall (6). The left thumb is outside the wall and the other fingers inside pull toward the thumb. This wall is raised and lowered two or three times to provide evenness and strength. Then the mass is compressed from the outside with both hands. From this point, large dishes, jars, bowls, and vases are thrown.

Large plate or dish (fig. 59). The preparatory steps given above are followed with adjustments for intended size. With the left hand on

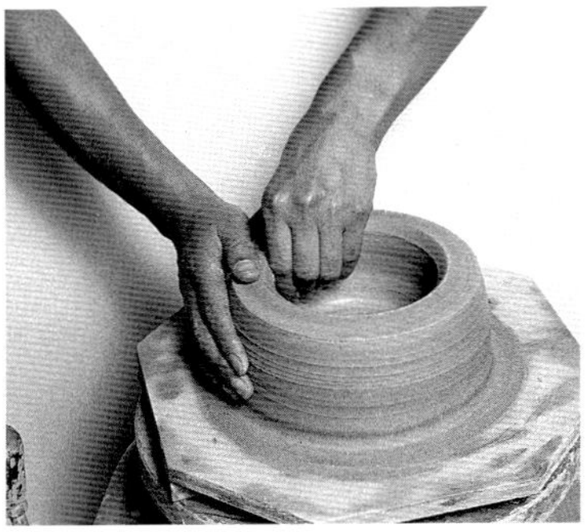

1

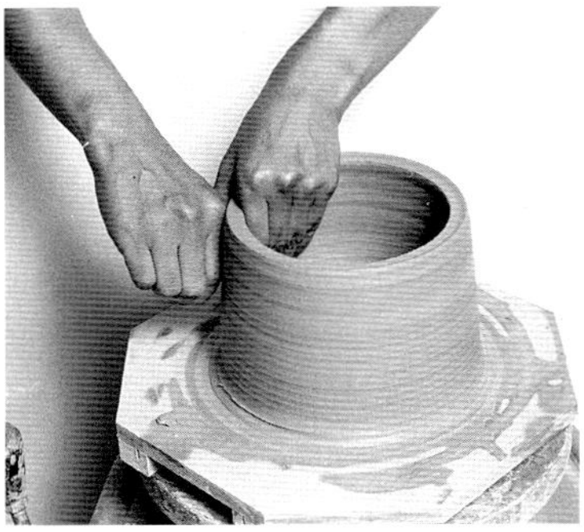

2

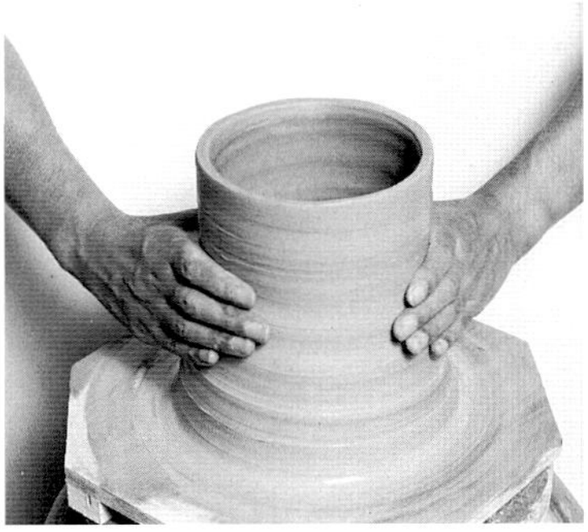

3

60. Large Jar

the inside and the right hand on the outside, a wall is pulled outward and upward to a point a little higher than the final intended height (1). The potter takes care not to let the rim get too thin and not to make accidental depressions in the center. The wheel is slowed to about half its original speed and the forming rib is applied, moving from the center outwards, with the right hand reinforcing from the outside (2). The plate is gradually made thinner, wider, and flatter. In this stage, it is especially important not to form a depression on the inside contour where the supported base stops and the unsupported wall begins. After achieving the desired shape, the potter cleans the inside contour with a finishing rib and smooths the rim with the chamois (3). A groove is incised around the base of the pot, and the cutting wire is run underneath (4). The bat is gently pried off the wheelhead using a stick (5).

Large jar (fig. 60). The coil-and-paddle technique for a large jar was mentioned previously, but large jars up to a certain size are also fashioned entirely on the wheel, usually on top of a bat. The preparatory steps are the same as for the large plate. With the left hand on the inside and the right hand on the outside, the potter pulls the clay straight upwards. In the early stages the left hand pulls the clay up, and the fingers and the palm of the right hand reinforce the outer wall (1). The wheel moves at a moderate speed, about 40 revolutions per minute. It takes three or four pulls to get the wall up(2). The potter is careful not to let the lower wall get too thin or to let the clay twist on account of too much pressure. If the upper wall starts flaring out, the rim is compressed from the outside(3). At this stage, the diameter of the rim is kept smaller than the diameter of the base.

As the wall approaches its final thickness and height (4), the shaping begins. With the index finger and possibly other fingers of

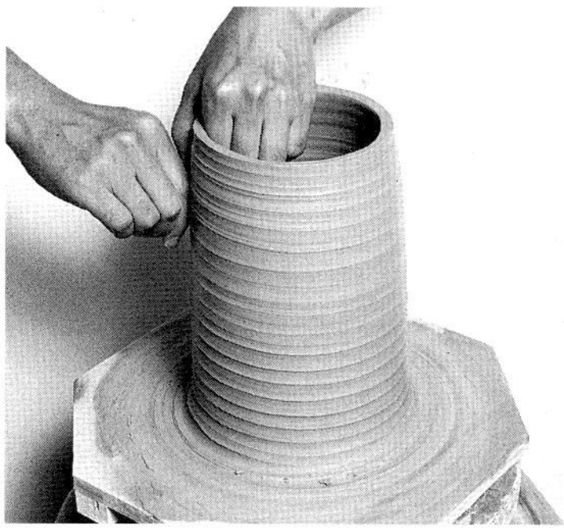

4

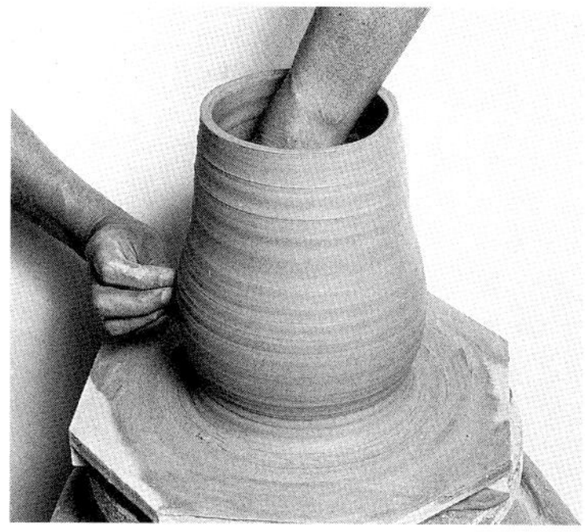

5

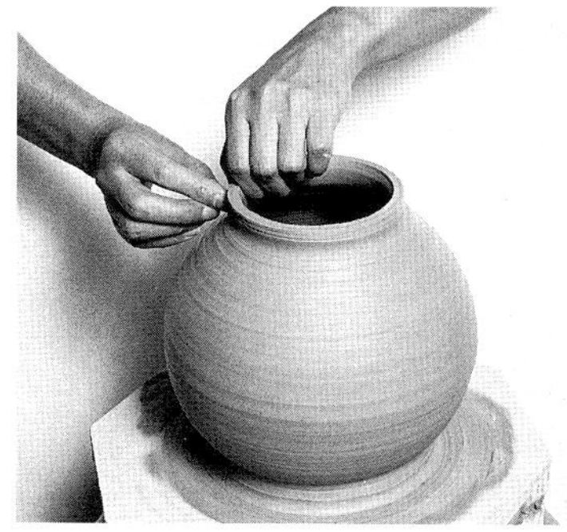

6

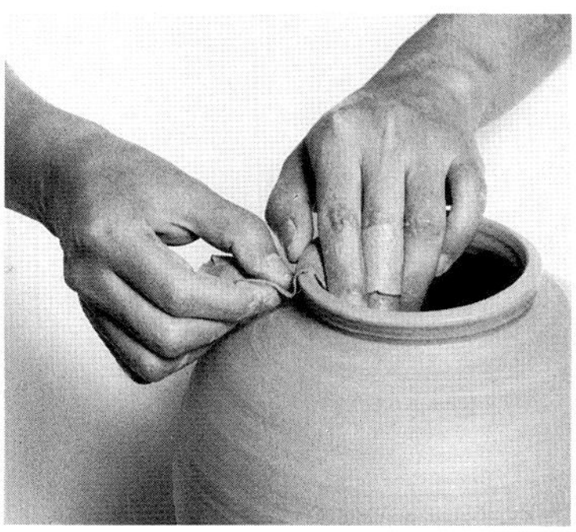

7

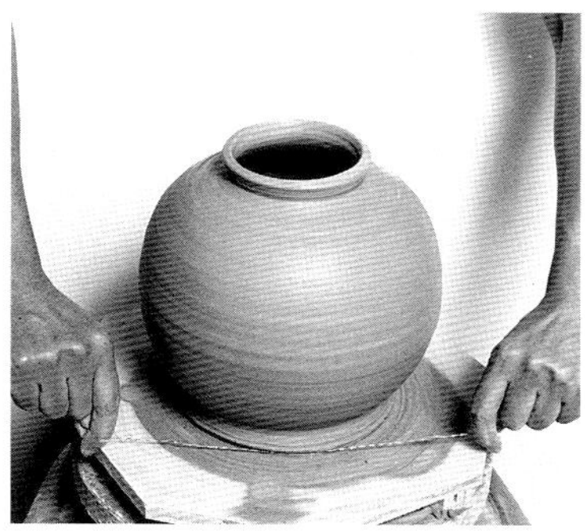

8

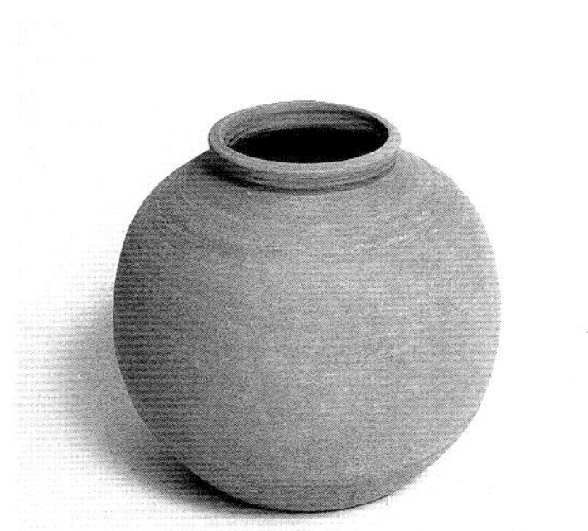

9

the left hand touching the inside and four fingertips of the right hand touching the outside, the hands simultaneously compress the wall and guide its swell (5). The throwing stick is used when necessary. At this stage the wheel moves more slowly, about 20 to 30 revolutions per minute. The work proceeds with deliberation, but too much manipulation reduces the plasticity of the clay and weakens the design. When the desired form is achieved, the rim is shaped (6). The contour should be strong and decisive, since the rim tends to dominate the appearance of the entire pot. After that, the rim is smoothed with the chamois (7). A groove is incised around the base, the cutting wire is run under the pot (8), and pot and bat are removed from the wheelhead. The pot is now finished (9).

In certain kilns in northern Japan, large vertical forms are made by inserting a throwing cylinder, a tall wooden cylinder somewhat narrower than the intended form, into a partially thrown pot. The walls are then raised by an external "choking" pressure from both hands.

61. Spouted bowl

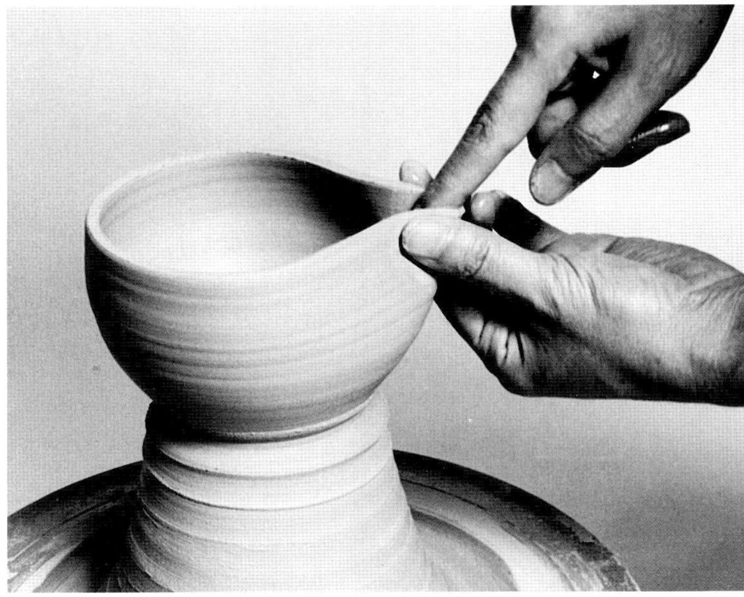

ALTERING THROWN FORMS

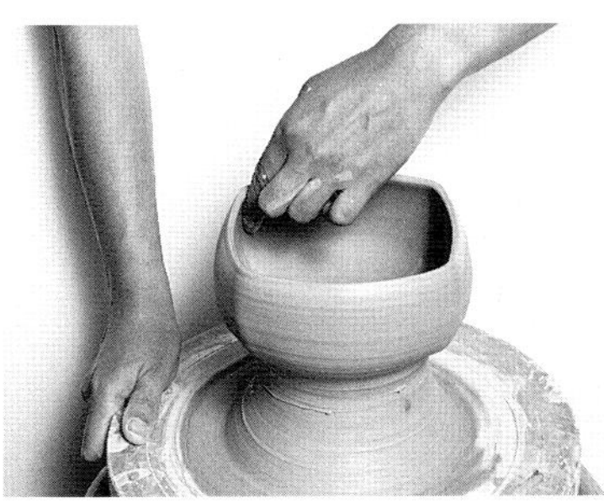

1

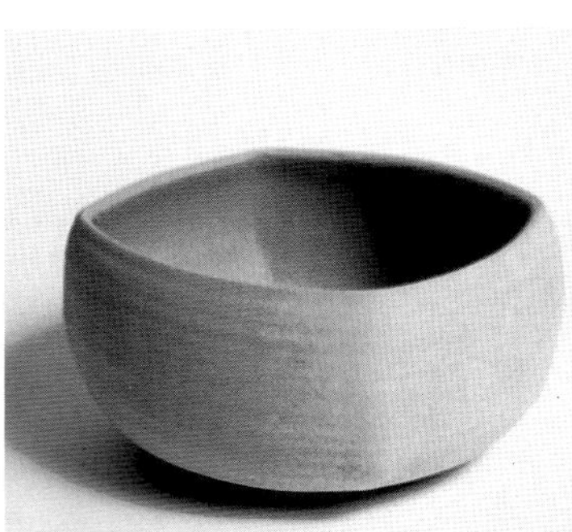

2

62. Cornered vessel

Thrown pots do not have to remain round. Some of the greatest traditional designs came from simple manipulations of thrown forms. For example, the Karatsu potters made endless permutations of dishes and bowls by bending edges, turning up corners, and poking indentations with the fingers. Modifications may be introduced when the piece is still on the wheel, as well as in the drying stages, by paddling, carving, or additions of parts.

Alteration in Plastic State

Spouted bowl (fig. 61). A cup or small bowl is made on the wheel according to the instructions given previously. Before the form is cut off the clay mass, a spout is pulled out with the finger of one hand dragging though the rim and the thumb and index finger of the other hand supporting from below. A large spout is formed with two fingers and a smaller spout with a thin tool handle. The more the edge of the spout is thinned, the less it will drip. The trimming of this object is the same as for an ordinary bowl. Care is taken that the pot is at true center and that the tools or hands do not catch on the spout.

Cornered vessel (fig. 62). A bowl is made on the wheel using a plastic clay. After finishing the rim with the chamois, the index finger pulls upward and outward at each corner of the bowl. The wheel is

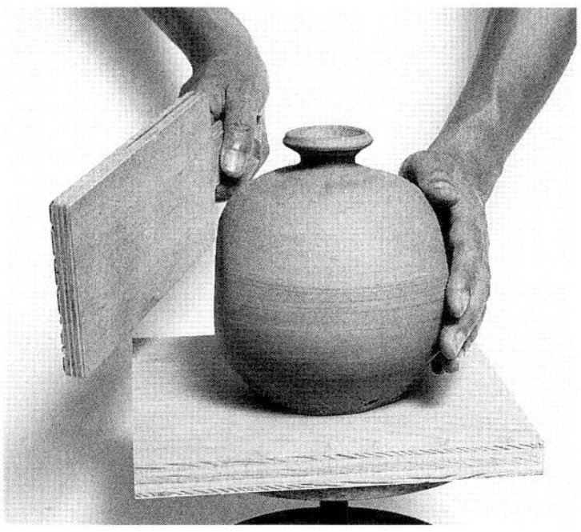

1

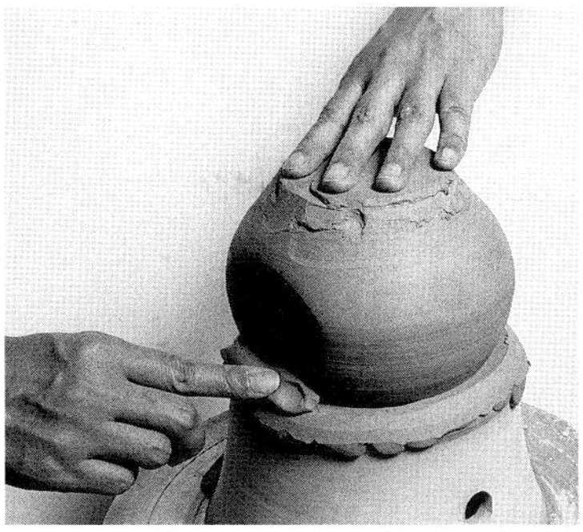

2

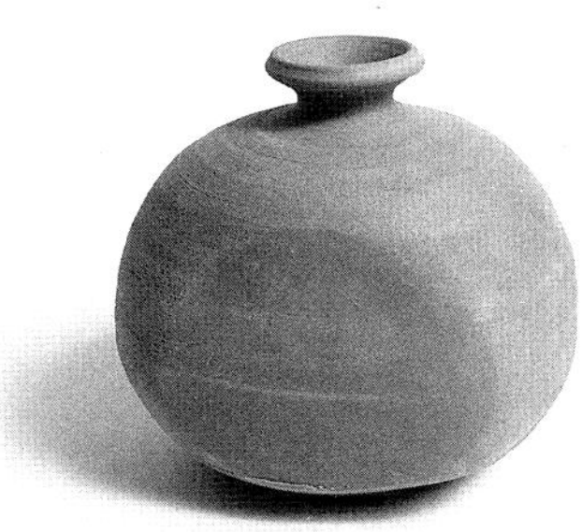

3

63. Paddled bottle

held in place with the other hand. The alteration is finished by tugging at each corner with the index finger (1). For trimming it is necessary to use a chuck (see trimming section, below) that engages the bottom of the piece rather than the sides, which tend to be weak. The finished pot is shown in (2).

Paddled bottle (fig. 63). A common way to transform a spherical shape is through paddling. In this case the form is modified after being taken off the wheel. A bottle form is made and dried to near leather hardness. Before paddling, either the paddle or the vessel is dusted with starch. The surface is paddled gently, with care taken that cracks do not form in the clay (1). For trimming (discussed further below), a chuck that will accommodate the upper part of the bottle is selected and the bottle is inserted upside down. The wheel is revolved to check that the bottle is centered. In order to support the paddled surfaces that are not flush with the chuck, gaps are filled in with coils of clay (2). The trimming is done slowly, and further paddling may be done afterwards. The finished pot is shown in (3). Note that this technique works better with vessels that have a small mouth. The greater the diameter of the vessel's mouth, the more it will be affected by the paddling.

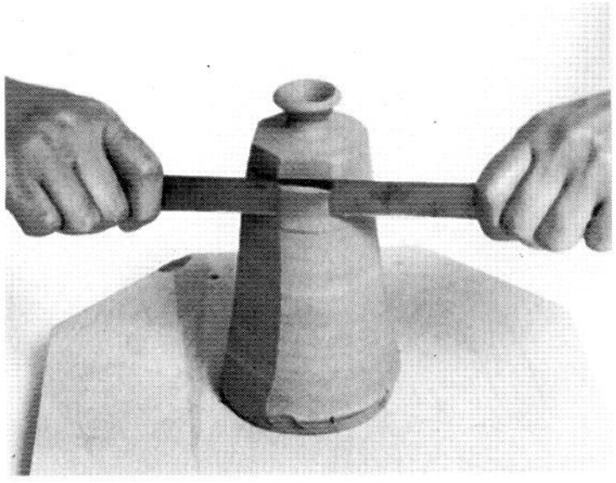

1

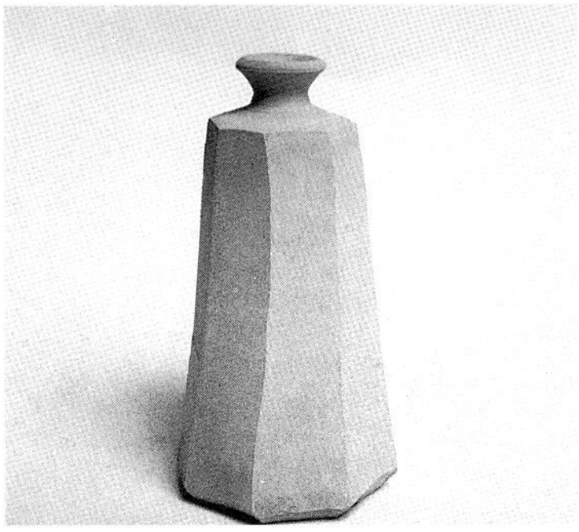

2

64. Faceting

Faceting (fig. 64). A vessel is made with somewhat thicker walls than usual and dried to leather hardness. Cuts are made in the wall of the vessel, following its contour (1). One may either divide and mark the segments in advance for an even effect or work spontaneously for an irregular effect. If the faceting extends down to the base of the pot, the edges of the base are beveled to avoid chipping. The finished pot is shown in (2). It is also possible to cut narrower facets after the pot has been trimmed.

Addition of Parts

The teapot, which involves complex problems of shape, composition, and assembly, is the most difficult of the traditional forms. There are two principal variations. The *dobin* (see pl. 20) is the larger form, used for large servings and multiple infusions of tea. It is identified by the cane or ceramic handle that arches over the lid. The side-handled teapot *(kyūsu),* usually smaller in size, is for small servings of tea (see pl. 19). The handle is ceramic, attached to the side of the vessel. With the exception of the handles, the procedures for the *dobin* and *kyūsu* are the same. Since others have described the making of the *dobin,* here the focus will be on the *kyūsu* shape, which can also be adapted to serve as a sauce or dressing server.

Side-handled teapot (fig. 65). The body, lid, spout, strainer, and handle of the teapot are all made separately and then assembled. The body is made in the same way as that of the teacup or small bottle. After sectioning out the requisite amount of clay from the clay mass, the left thumb is sunk into the top to form a well. The walls are pulled out and upward. Extra clay is left around the rim to fashion a seating for the lid. Since more forming will have to be done after shaping the seating, the lower walls are left a little bit thicker than usual. After the basic shape is made, the seating is formed. With the index finger of the left hand on the top of the rim, the left middle finger providing inside support, and the right index finger providing outside support, the left index finger presses down (1). This press originates from the inner edge of the rim and moves in a downward and slightly outward direction. The outer edge of the rim rises, forming a kind of step. This is the seating. Then the body is given its final shaping, with care taken not to alter the seating (2). The seating diameter is measured with calipers or a dragonfly gauge (3). The rim is finished with a wet chamois, and the body is cut off and set aside to dry.

For the handle, a cylinder about 3 centimeters wide is fashioned on the top of the hump. The left thumb is inserted at the top and a deep hole is made. A narrow cylinder is pulled up, using the throwing stick if necessary (4).

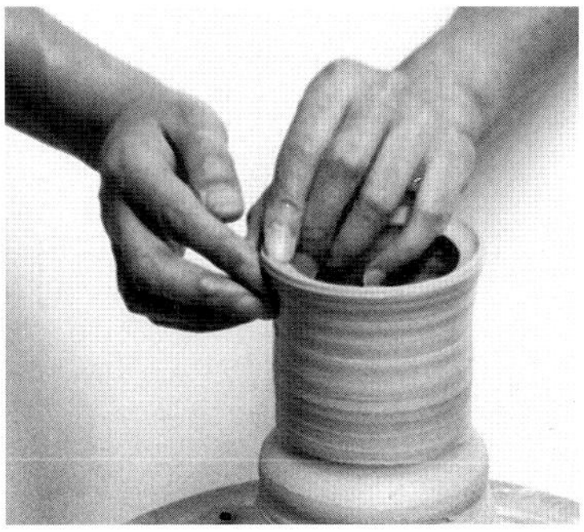

1

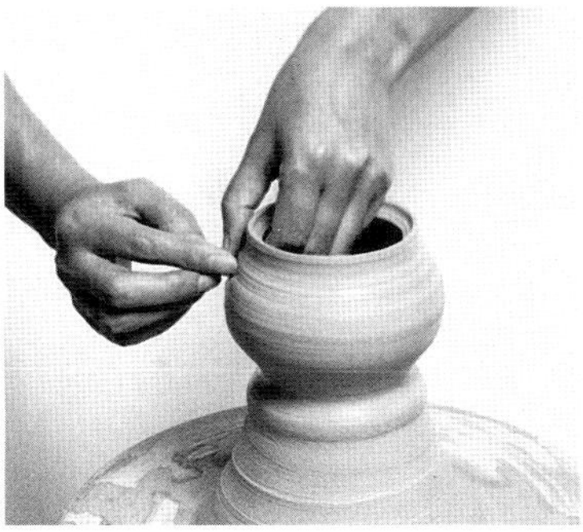

2

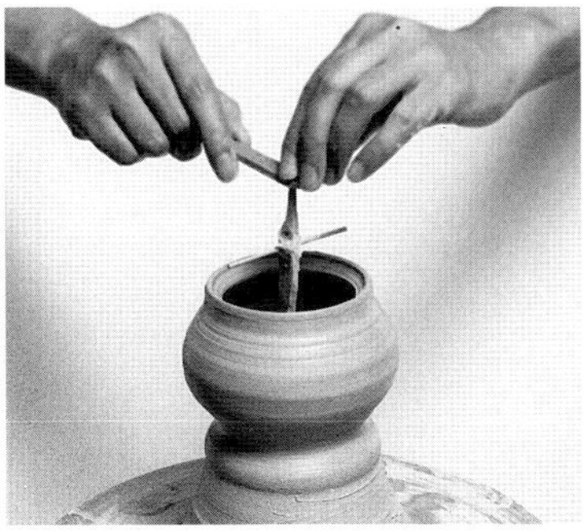

3

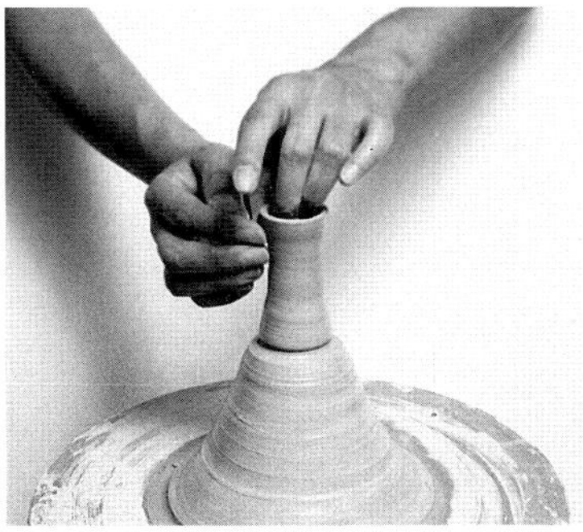

4

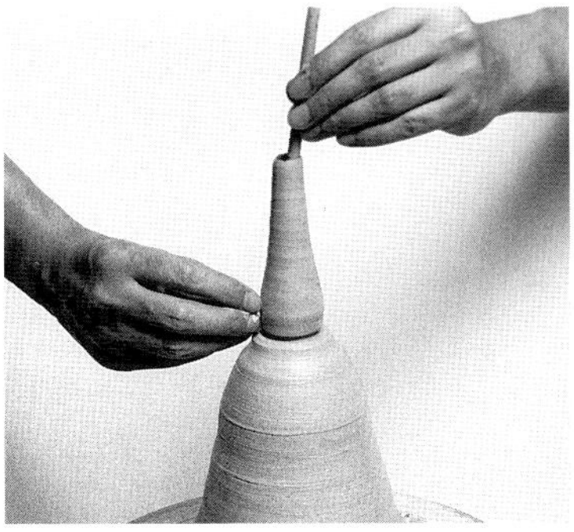

5

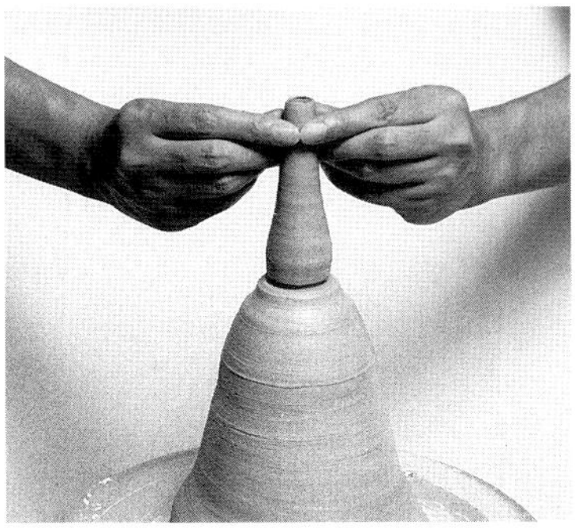

6

65. Side-handled teapot
(Continued on next page)

For the spout, another small cylinder is thrown, even narrower than the one used for the handle. Since the finger will not fit inside the cylinder, a narrow tool is used in the same way that the throwing stick is used in making bottles (5). The walls should be as thin as possible, and the diameter of the mouth should be about 1 centimeter.

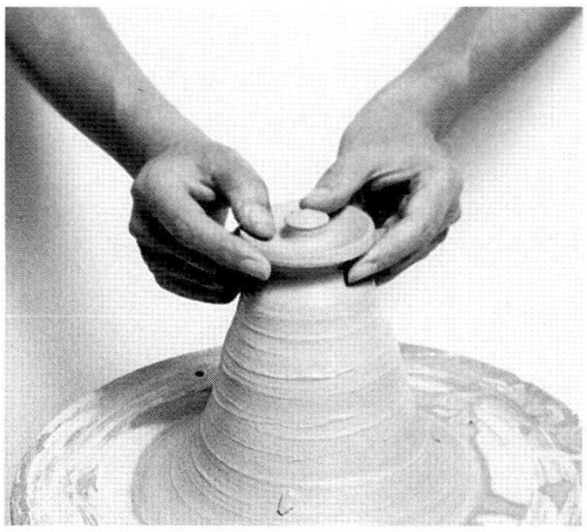

7

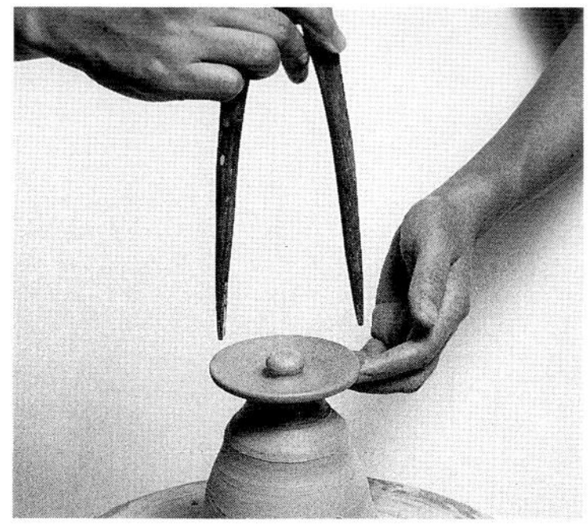

8

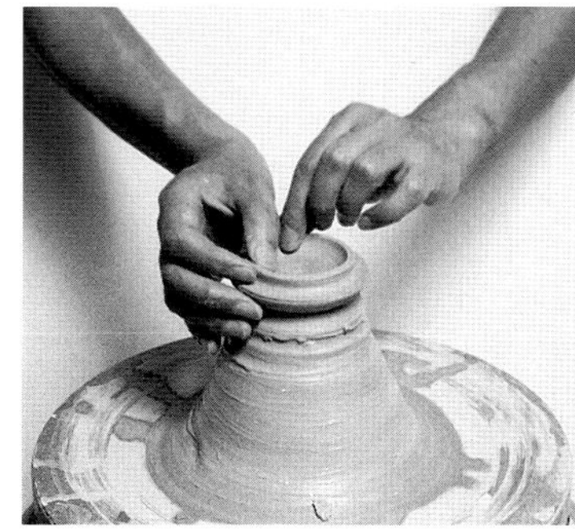

9

65. Side-handled teapot (Continued)

A thin spout can be obtained by alternately pulling up the cylinder through throwing and then consolidating it by collaring it with the fingers of both hands (6).

There are two ways of making the lid of the teapot. The first, a concave lid, is for a body with a seating. The basic throwing sequence for a small plate is followed, but the horizontal pull is started slightly off center, which will leave a small knob in the middle. A stem is formed under the knob by gently pressing down and inward (7). The knob is shaped. A good knob should have some formal relationship to the overall teapot shape, and moreover it should be easy to grasp. After the knob has been formed, the shape of the lid is finished. The lid diameter is measured with calipers (8). At this stage the lid is left a little oversized—it can always be trimmed down.

The second method, for a convex lid, usually accompanies a body with no seating. The lid is thrown with a flange, and the knob is trimmed out from the bottom or attached later. A disc of clay is sectioned out on top of the hump and the clay is pulled out into a shallow plate form using the basic throwing procedure. Some extra thickness is left at the rim. With the inside reinforced by the right thumb and the top of the rim consolidated by the right middle finger, the right index finger presses inward at the rim. This causes a flange to rise (9). The flange is shaped with pressure from the left thumb and index fingers. The diameter is checked with the calipers (10), and the rim is finished with a wet chamois.

The strainer functions to keep the tea leaves inside the teapot, and just punching holes into the wall of the body will fulfill that purpose. But a convex strainer inside the body has a larger surface, with more holes, allowing a greater volume of liquid into the spout. The convex shape also helps to keep the tea leaves from plugging up the

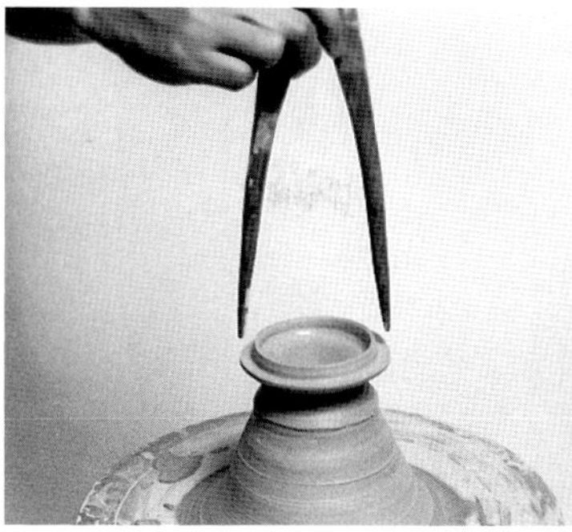

10

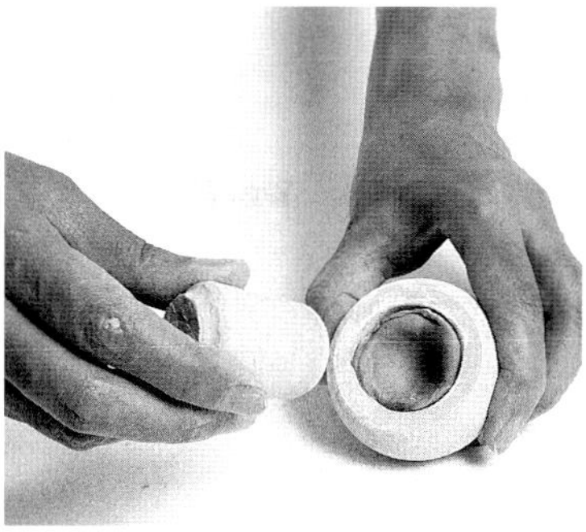

11

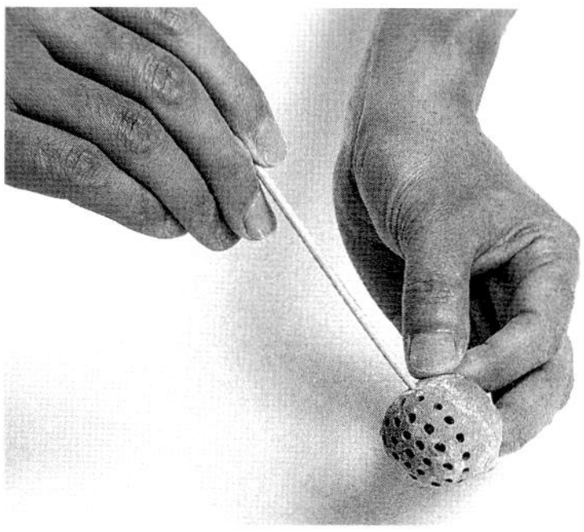

12

holes. A small cup shape can be pinched out, but for a more uniform shape, a two-piece mold and matching press is made of bisque-fired clay or plaster and a thin slab of clay is pressed into it (11). This is dried to leather hardness. To make the holes, the cup is perforated with a sharpened umbrella spoke or similar object as the fingers support it from inside (12).

The body is trimmed according to the instructions in the trimming section, although teapots generally have recessed bases rather than projecting foot rings. Depending on its shape, the rim may either be inserted inside or fitted over the trimming chuck. Since added parts like the handle and spout will affect the total balance of the finished piece, the potter takes care that the trimmed foot is sufficiently broad and stable.

The concave lid with a thrown knob may be trimmed by hand or on the wheel, but if the original cutoff is neat and close enough to the lid, trimming may not be necessary at all. When trimming the bottom of the lid, the potter follows the outer contours closely. The fit between lid and body are checked: if the lid is too large, its edge is trimmed. Shrinkage differential between lid and body must be anticipated. The rim of the lid is finished with a moistened chamois.

The convex lid has no knob, so it must be created in the trimming stage. A chuck smaller in diameter than the flange is readied and the lid is placed over it. The top is first trimmed level, and then excess clay is trimmed from the perimeter, leaving the projecting knob. The side of the knob is beveled and trimmed (13). The contour is finished, the fit with the body is checked, and necessary adjustments are made.

The body, strainer, spout, handle, and lid, five parts in all, are gathered together (14). The spout is separated from its thick base

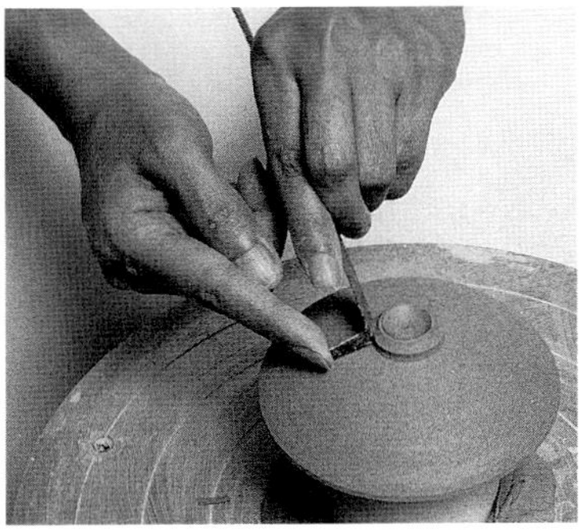

13

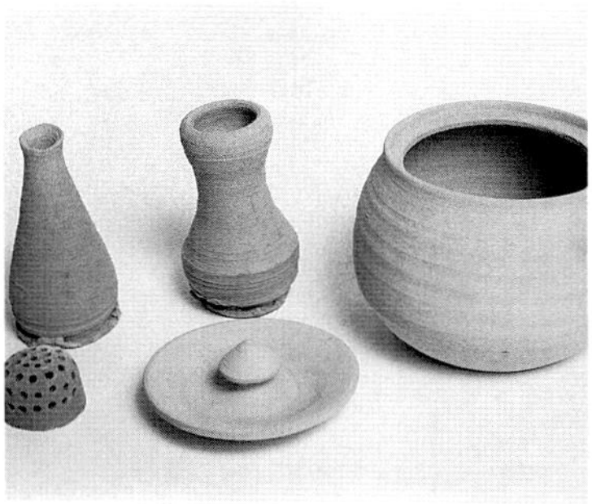

14

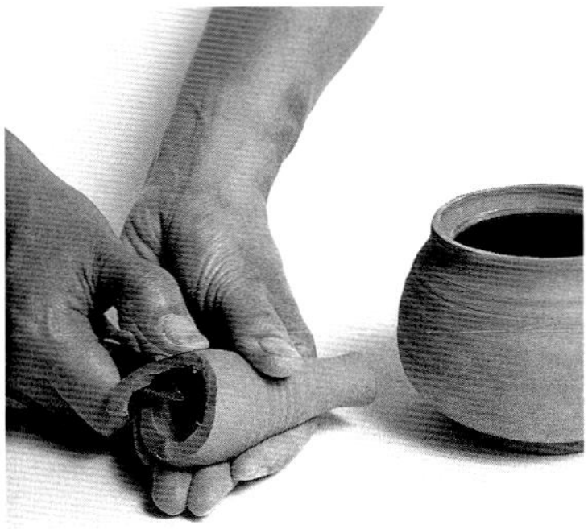

15

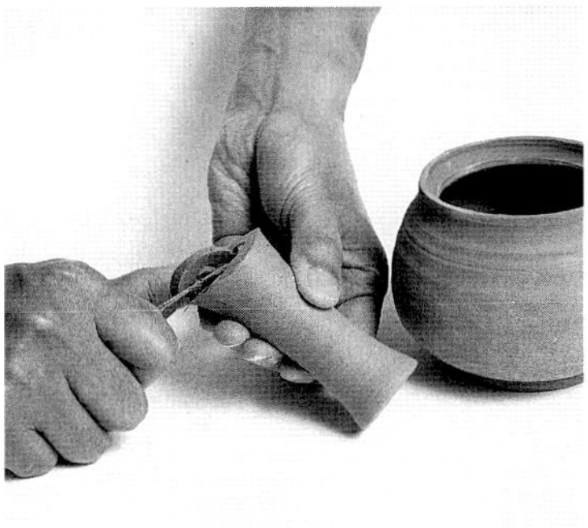

16

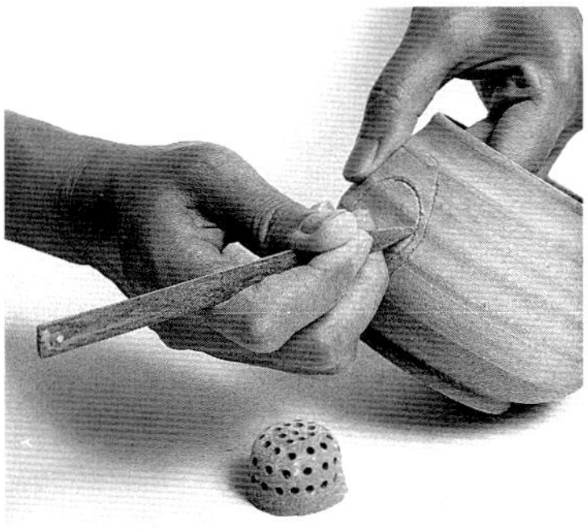

17

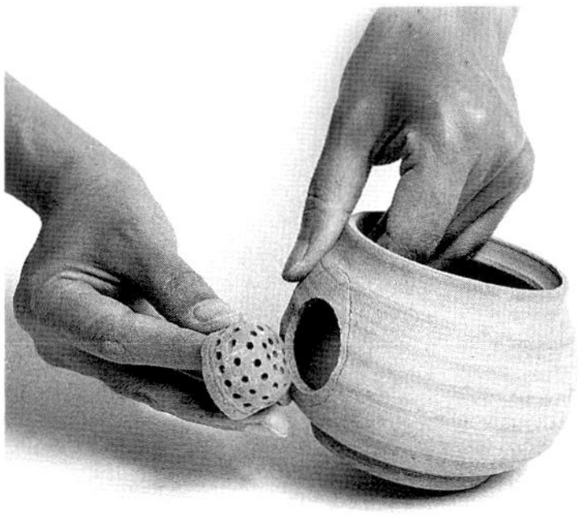

18

65. Side-handled teapot (Continued)

with a diagonal cut, and its base is contoured to match its intended position on the body (15). Ideally, the bottom edge of the spout mouth should be the same height as the mouth rim of the body. If the top of the spout is too low, the liquid will pour out before the pot is full. The base of the handle is contoured in the same manner as that of the spout (16).

With the spout held against the desired location on the body, its base is outlined on the body. A circle is traced inside this outline (17). If a strainer is not used, holes are drilled into the body in the area of the circle with a sharpened umbrella spoke at this stage. If a strainer is used, the circle is cut out. The edges of the cut should bevel outward to help the strainer fit into the hole. The strainer is gently pushed into the hole (18). The edges are luted, and the inside is inspected for a tight fit (19). The base of the spout and the place where it will fit against the body are moistened and scored, and then slip is applied to the edges (20). The spout is attached, and the joint is cleaned up with water or slip (21). The handle is attached in the

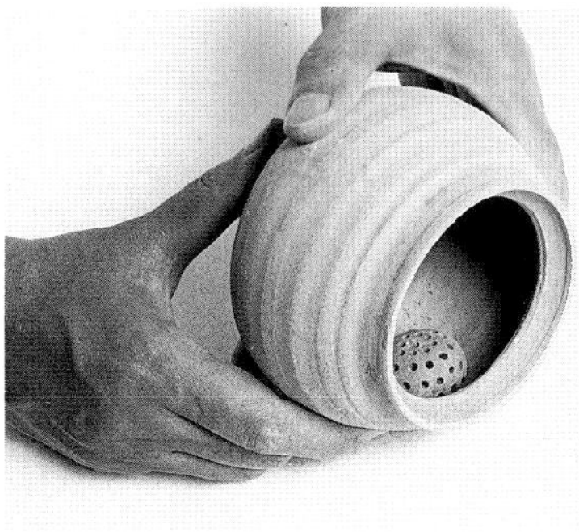

19

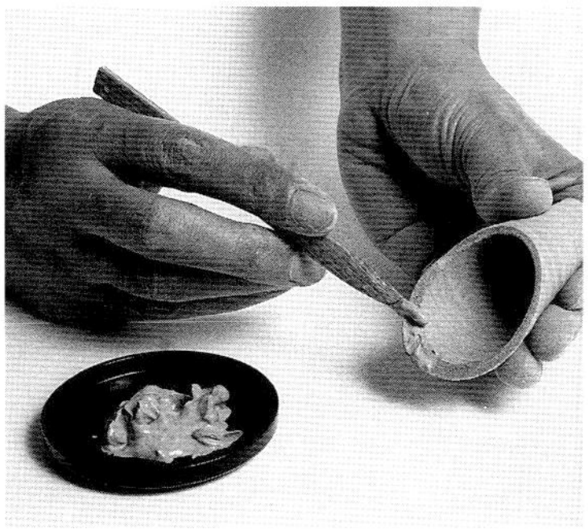

20

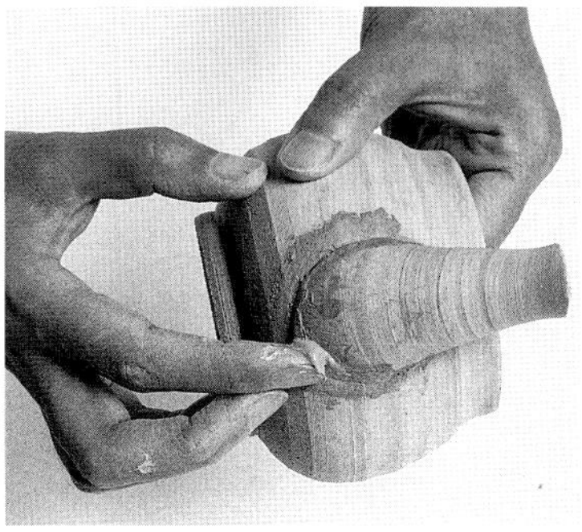

21

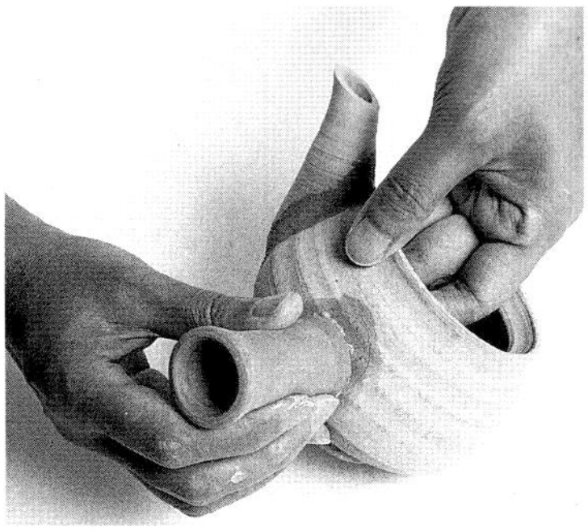

22

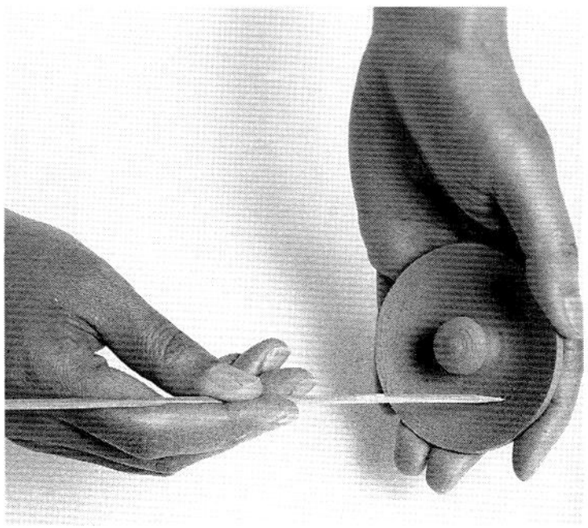

23

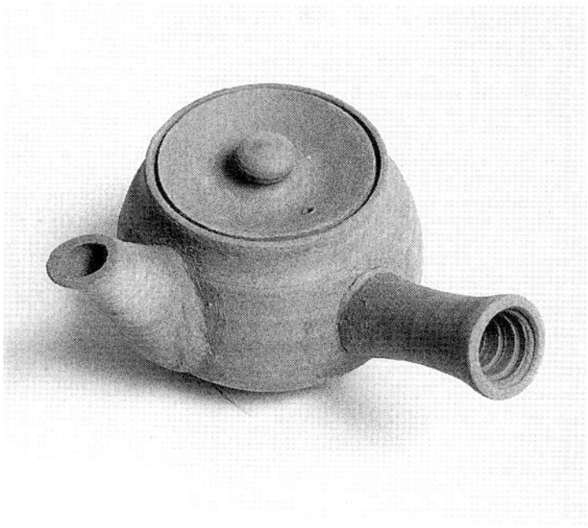

24

same manner (22). When viewed from above, the axis of the spout is at an eighty degree angle to the axis of the handle, with the spout pouring to the left (a concession to right-handed users). A small hole is drilled in the lid to relieve air pressure during pouring (23). The finished pot is shown in (24).

TRIMMING

In designing pots, it is always worthwhile to consider whether they need any trimming. With careful throwing, some shaving of the lower walls of a vessel, and a skillful severing of the pot from the clay mass, some types of wares, especially those that are not moved during use, need no trimming. Indeed, in certain pots the spiral-shaped mark left by the cutting cord is seen as a sign of aesthetic merit.

Most pots for table use are trimmed, however, because a trimmed foot has certain functional and aesthetic advantages. The

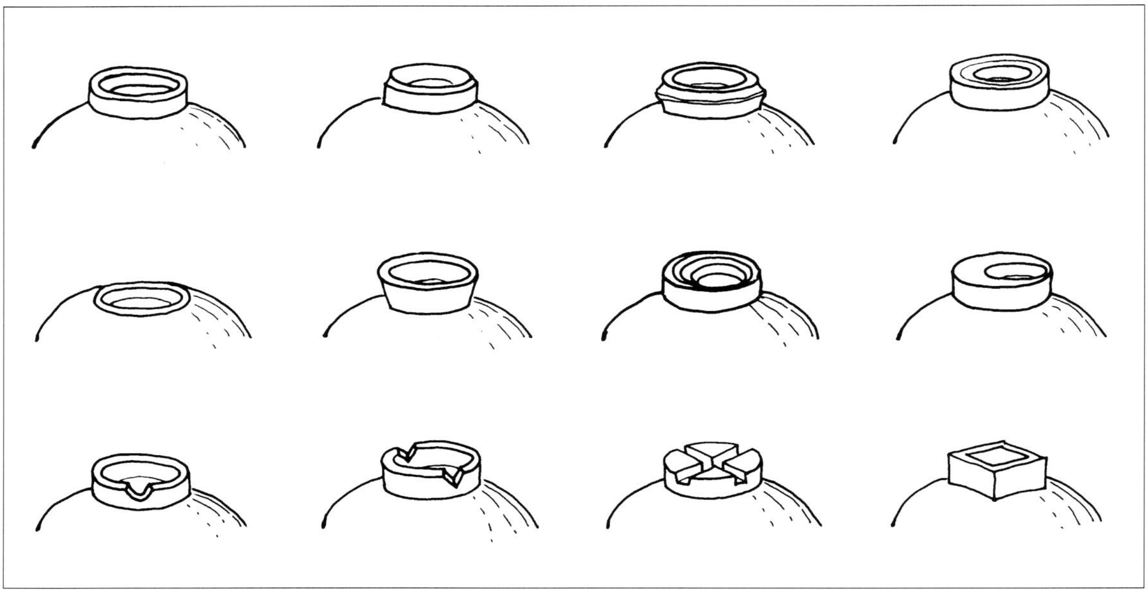

66. Types of foot ring

foot gives physical stability to the pot and minimizes damage to the surface that the pot comes into contact with. Since most traditional Japanese pots do not have handles, the foot makes it possible to slip the fingers under the pot for lifting. As mentioned in Chapter 1, the foot ring is also seen as the place where the potter reveals his or her "naked" craft. Poetic names for certain foot ring shapes—"bamboo joint," "half moon," "plectrum"—attest to the love of this part of the pot. In addition to the many possible designs for a foot (fig. 66), the nature of the trimmed surface is also a focal point. For example, the shredded or "crepe" surface on the foot ring of Karatsu ware has long been admired (see fig 134). This effect appears when a sandy clay is trimmed with a bamboo or wooden spatula after a minimal drying time. In Karatsu as well as in other Kyushu kilns, the trimmer's wheel moves counterclockwise to the thrower's clockwise, which turns up more texture in the clay.

Aside from the stylistic considerations, a good foot ring has what the rest of a good pot has, a sharp sense of what is appropriate for the material and the function of the piece. Mindful of the fact that excessive trimming tends to weaken the appearance of a pot, throwers leave as little excess clay at the bottom as possible, so that the job can be accomplished quickly and decisively, although there are exceptions such as overall trimming for textural purposes or for exceptionally thin walls. Timing of the trimming is crucial. If the pot is too moist, it will deform in trimming and the clay will stick to the tools. If it is too hard, the knife will chatter on its surface. The angle at which the knife is held and the speed of the wheel are also critical. Knives are kept free of rust and are sharpened frequently. Old pottery manuals caution potters to heat their knives before sharpening them.

Japanese potters rarely trim small- to medium-size vessels directly on the wheelhead. Instead they mount the ware on or in chucks (for diagrams, see p. 39). Chucks are open-ended cylinders, usually with a hole on the side to let out air pressure. The size and shape of the chuck are tailored to the vessel in question: narrow cylinders of medium height for teacups and small bowls; broad, low cylinders for large dishes; and deep cylinders for bottles and jars. The diameter of the chuck varies widely, but for a tea bowl or cup it is about half the diameter of the vessel itself. Special attention is paid to the bevel on the chuck rim, where it contacts the pot. Potters who work in limited production make a new chuck with every batch of pots: at the end of the throwing cycle, clay is left over on the wheel-head and a chuck is fashioned with that. Some potters use dry-hard chucks, briefly wetting both ends for adhesion before use. Plaster chucks are also frequently used. The most common type, however, is the bisque-fired chuck.

Preparing a bisque-fired chuck (fig. 67). Both ends of the chuck are briefly immersed in water. A coil of unfired clay of even thickness is applied around the upper rim (1). Another coil is laid in a ring on the wheelhead (2). The chuck is placed inside the ring and centered, then secured to the wheelhead (3). The wheel is started and the coil on the top of the chuck is trimmed smooth, with bevels inside and out (4). If the pot surfaces are extremely sensitive, a sheet of tissue paper is draped over the chuck before the pot is mounted.

Trimming a small cup (fig. 68). When the cup has reached sufficient hardness, it is placed over the chuck, and the wheel is started. With the wheel turning between 30 to 40 revolutions per minute, the base of the cup is tapped from the side until it is centered on the chuck (1). This is not easy to achieve at first: too harsh a blow will send the cup even further off kilter. But with a few gentle taps, the pot pops into center. The base is leveled first. A sharpened trimming

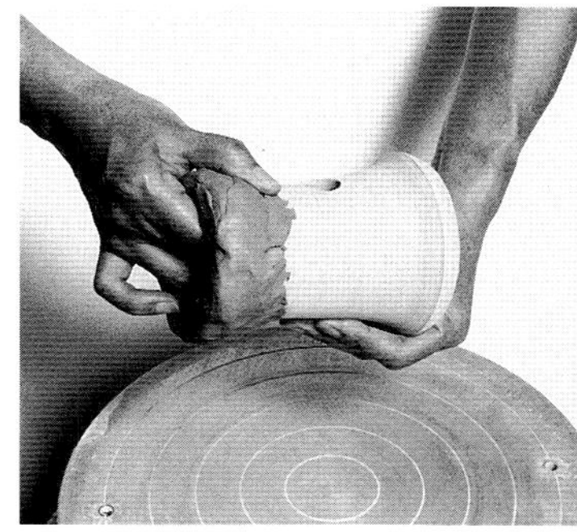
1

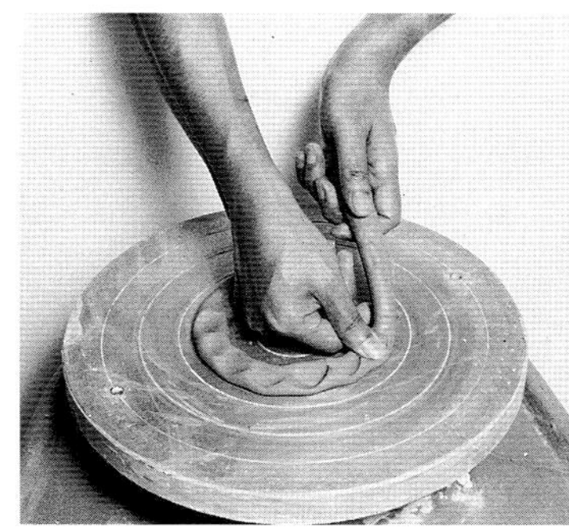
2

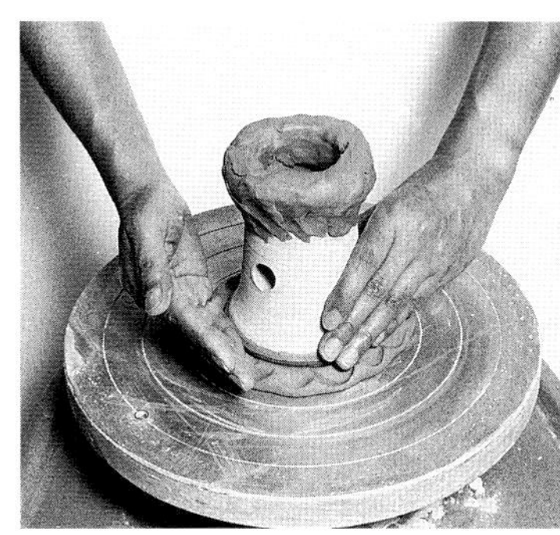
3

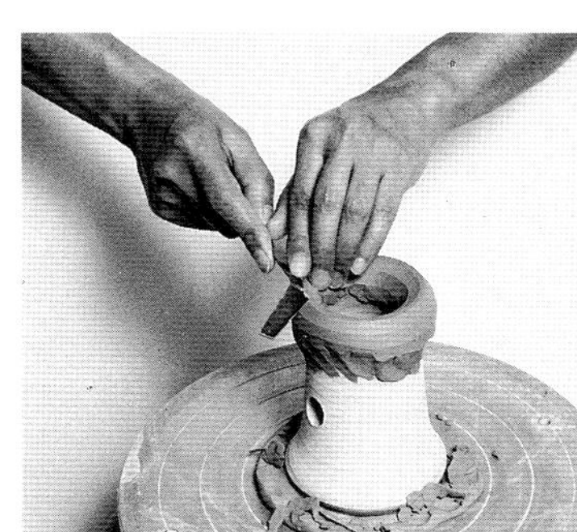
67. Preparing a chuck 4

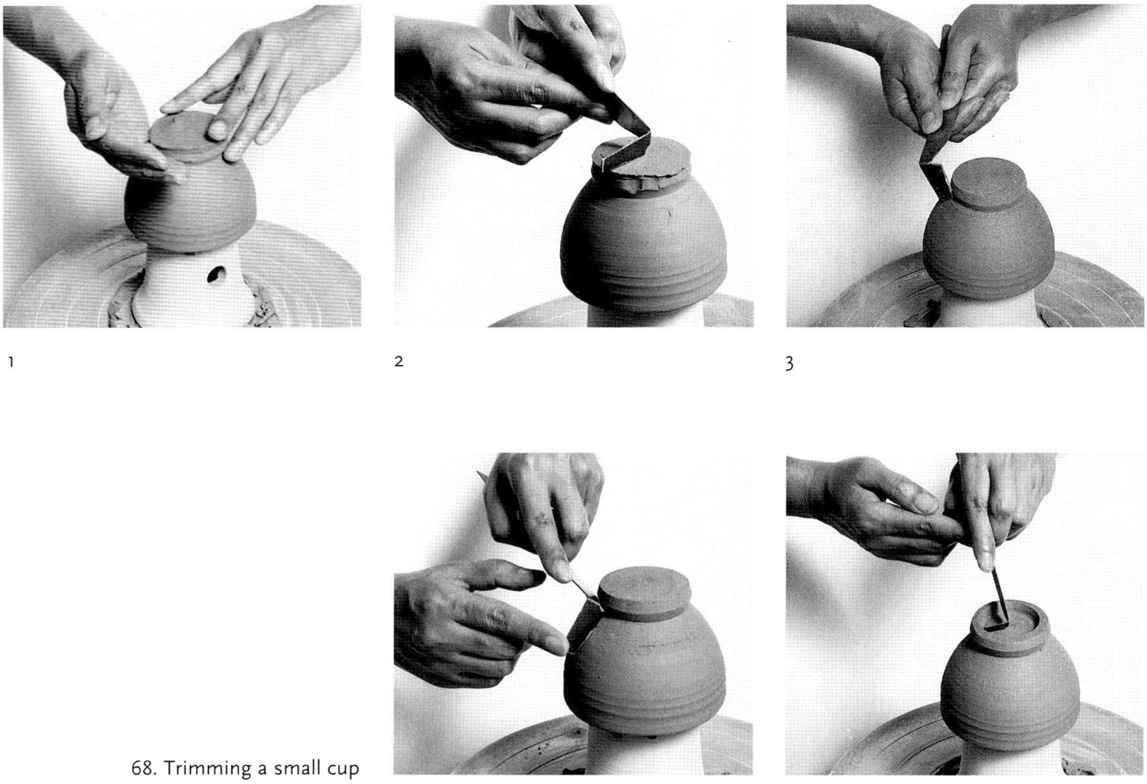

1 2 3 4 5

68. Trimming a small cup

knife is held like a pencil with the left hand and supported with the right. The angle between the tool and the surface is between 30 and 60 degrees. This angle is important and when maintained can make trimming much less of a task. The blade is engaged slowly (2). The base is first trimmed flat, starting in the center and working outward. The side of the base is trimmed, followed by the corner of the shoulder (3) and the lower wall of the vessel (4). For a foot ring or a recessed base, the inside of the base is trimmed (5). Close attention is paid to the thickness. If it is left too thick, the bottom will crack; if it is too thin, it will sag or even shatter in the firing. Potters working on large numbers of identical pieces rely on sight alone in estimating how much clay to take off, but in the case of individual pieces, the base is tapped with the finger. Uniform diameters can be gauged for the size of the foot by marking out a circle on the leveled base with a U-shaped compass fashioned from stiff wire. An adjustable pointer is sometimes used to mark uniform heights on untrimmed pieces. Trimmed pieces are placed bottom up on a paper-lined ware board and dried slowly in a shaded area.

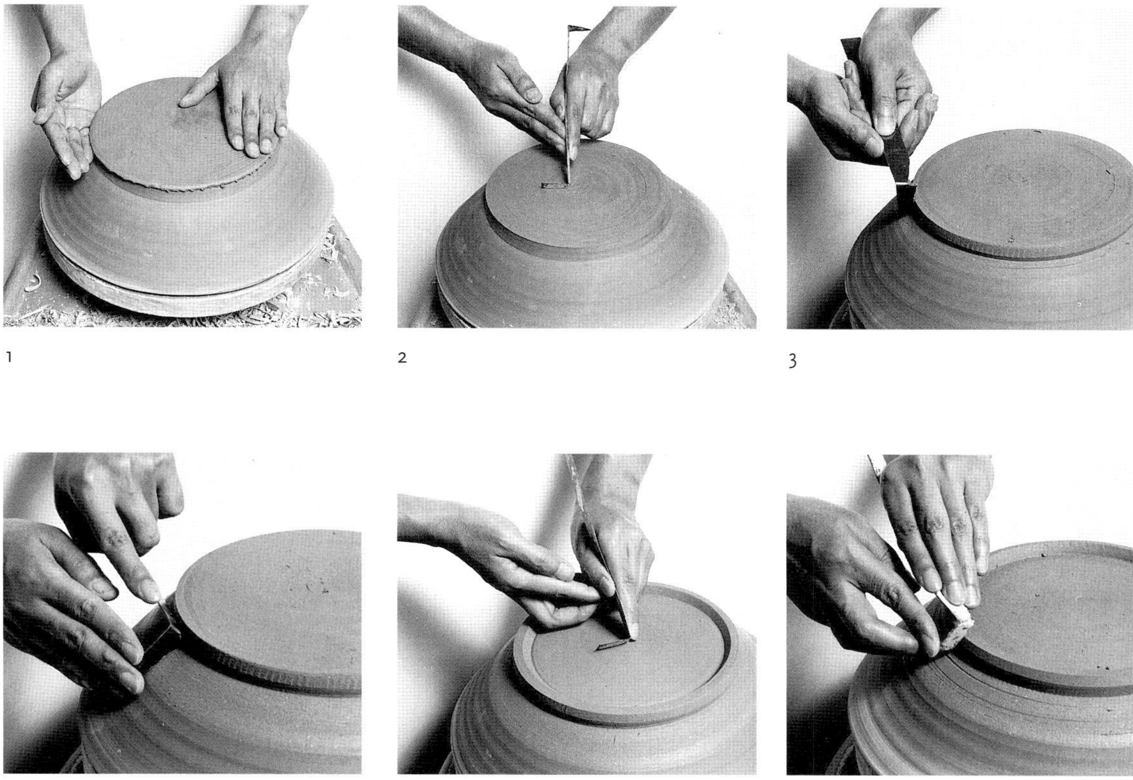

1 2 3 4 5 6

69. Trimming a large plate

If the pot rattles on the chuck in the course of trimming, it is a sign that the clay ring around the top of the chuck has become too dry and should be changed.

For trimming a small plate, the basic procedure is the same as for the cup, except that a wider chuck is used.

For trimming a small bottle or vase, the vessel is placed inside the chuck, so the latter must have some width (for the illustrations, see section on altering forms, p. 85). The chuck rim must have an inside bevel to protect the pot's walls.

Trimming a large plate, bowl, or dish (fig. 69). A low, wide chuck is centered on the wheelhead. The plate is placed over the chuck, the wheel is started at a slow speed, and the plate is centered (1). With the trimming knife in the left hand and the right hand supporting the piece, the potter trims the base flat. The blade, held at about a 30-degree angle to the surface, works from the center outward (2). After the size of the foot is determined, its side is trimmed (3). The wall is trimmed (4). The inside of the foot is trimmed, and is tapped, if necessary, to

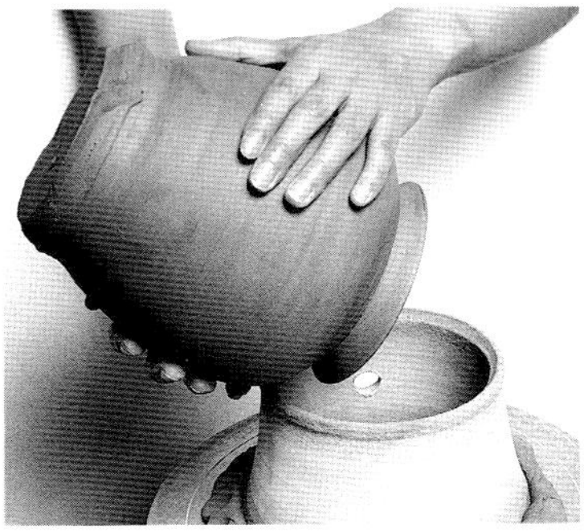

1

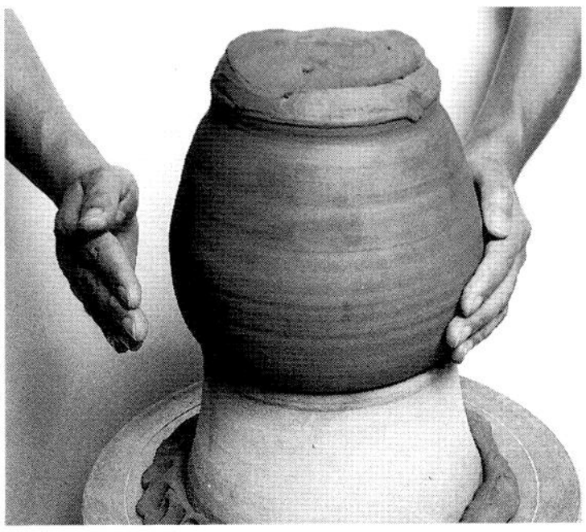

2

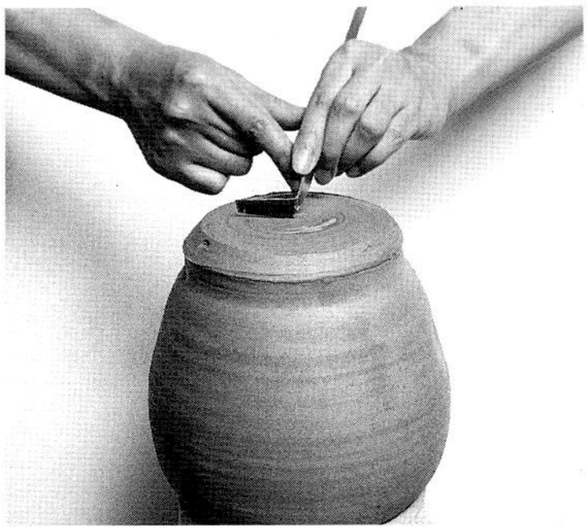

3

70. Trimming a large jar

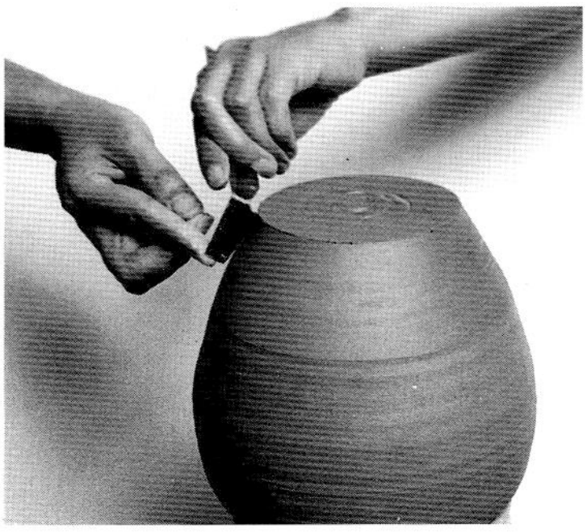

4

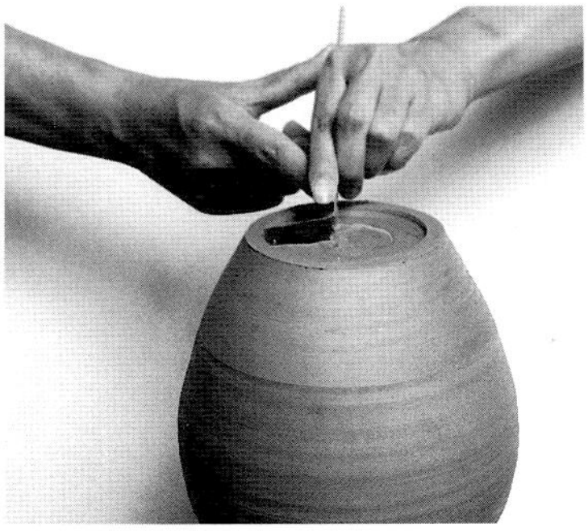

5

determine thickness (5). The foot ring is finished with appropriate cuts. The corners are usually beveled or smoothed to avoid chipping (6).

Trimming a large jar (fig. 70). The large jar is set into an appropriate chuck (1) and centered (2). The procedure is similar to that employed for the large plate, although the knife may be inclined more steeply. The bottom is trimmed (3), followed by the wall (4). Instead of the usual projecting foot ring, large jars usually have a flat or recessed foot (5).

A chuck is seldom used for wide, low forms such as large bowls. A piece of felt or an old blanket is cut out and centered on the wheelhead, and the pot is trimmed on top of that. It is necessary to trim slowly, for the cloth is easy to dislodge. Some potters lay down a slab of clay instead of cloth, but this tends to stick to the pot.

FOUR

Decoration

The prehistoric Japanese potter was a masterful decorator. Much of the early decoration, such as the dramatic reliefs on Jômon earthenwares, was of the plastic variety. As a result of the expanded continental influences beginning around the third century BCE, decoration became restrained, limited mostly to incised patterns. But with the development of transparent glazes and fine porcelain bodies in the seventeenth century, painted decoration became extremely popular. The earliest designs were simple and abstract, but the motifs gradually became more representational. By the mid-nineteenth century, the overwhelming demand for pictorial design caused potters to neglect the other facets of their craft.

This history reveals some of the options open to today's potter. The first and most important decision regards the relationship between the decoration and the vessel proper. If the potter aims for a type of decoration that creates a harmony with materials, form, and function, the traditional Japanese approach to decoration has much to offer. Japanese pottery decoration embodies several basic principles. One is that the decorating materials should be frankly ceramic, true to the nature of those materials rather than a simulation of other media such as plaster or oil paint. Second, decoration should complement the vessel form rather than obscure it: negative space is just as important as the positive, decorated surface. Third, graphic decoration should suggest through motif rather than represent through illusionistic picture. In the Japanese case, the power of suggestion is enhanced by an audience that appreciates the formal and suggestive qualities of calligraphic line. There is a wealth of traditional motifs to draw from, and new motifs can be derived from the observation of nature. Even complicated subjects reveal a simple underlying structure when studied. When this form combines with the decorator's own habits of gesture, new patterns emerge.

The Japanese potter has also found subjects and methods for decoration in tools, materials, and working rhythms, resulting in

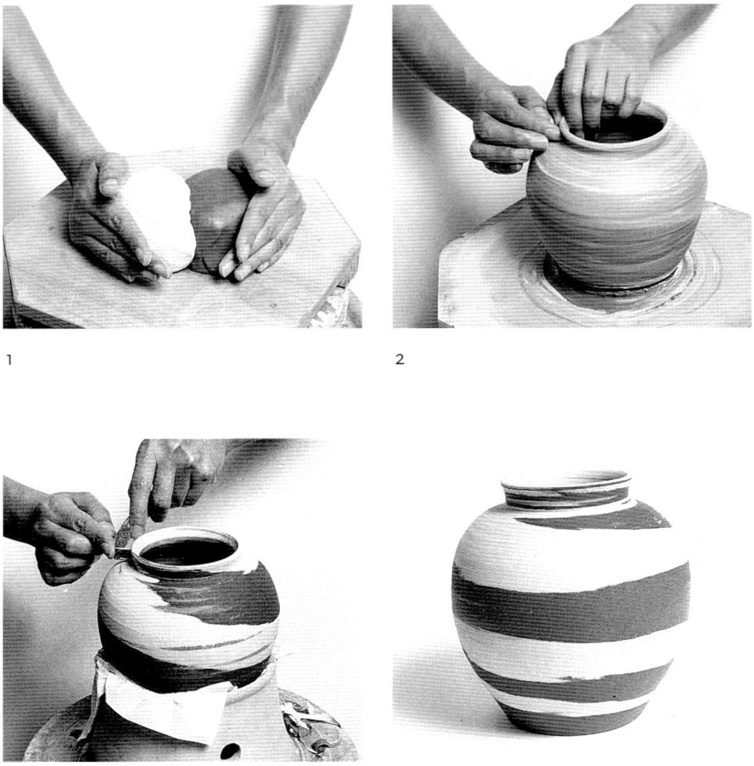

1 2 3 4

71. Marbling on the wheel

simple, spontaneous, and integral decorations. Slip decoration in particular epitomizes this approach, and for that reason it is given special attention in the following pages.

BEFORE BISQUE FIRING

Some of the ways in which pots can be changed in the wet state, such as faceting, are mentioned in the previous chapter.

Marbling

Marbling involves combining two or more clays of contasting color prior to forming to produce various decorative effects. The technique was developed in China in the eighth century. Although it appeared in Japan only in a few Shino-ware pieces from the late sixteenth century and in other regional products in the nineteenth century, it is now very popular. Historically, contrasts were obtained by pairing a white clay and an iron-bearing clay, but contemporary pot-

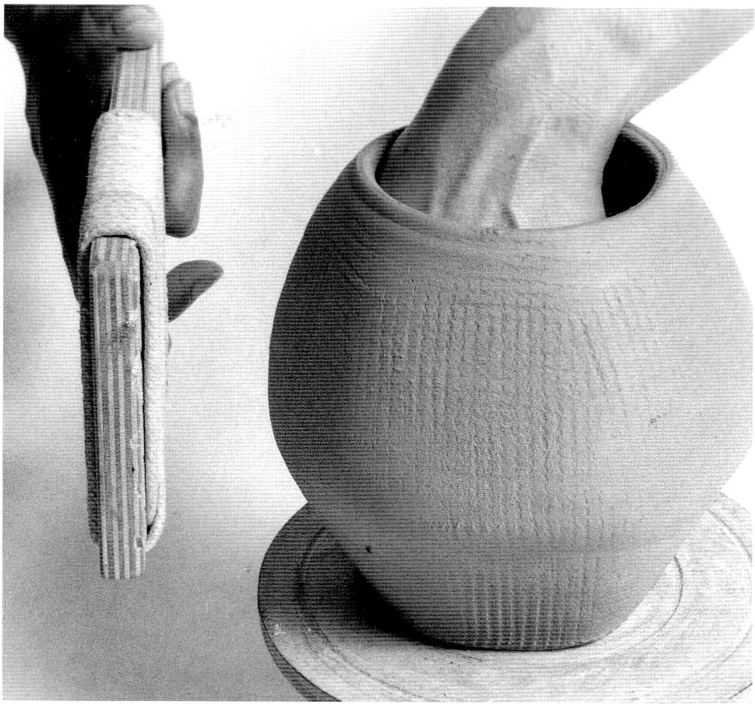

72. Texture paddling

ters often color their clays with refractory pigments of the spinel type. These come in a wide range of colors. Checkerboard and feather marbling are popular in Japan, but since they are well known to Western potters, wheel marbling, a traditional Japanese technique, is introduced here.

Marbling combined with forming (fig. 71). Two clays of contrasting color and identical consistency are slapped together on the wheelhead (1). While throwing the pot, the potter avoids the excessive manipulation that would homogenize the mass (2). After forming, the clay contrast is blurred at the surface level, so the surface of the pot is trimmed (3). The finished pot is shown in (4).

Paddling

This technique originated from the need to fuse the coils in coil-made pots. Smooth paddles stick to the pot, but textured paddles do not. These textured paddles leave marks on the clay, providing decorative possibilities. For example, after the form is made, the surface is paddled, with a hand or an anvil inside the pot providing a support (fig. 72). It is also possible to paddle a textured or patterned object into the wet clay (fig. 73).

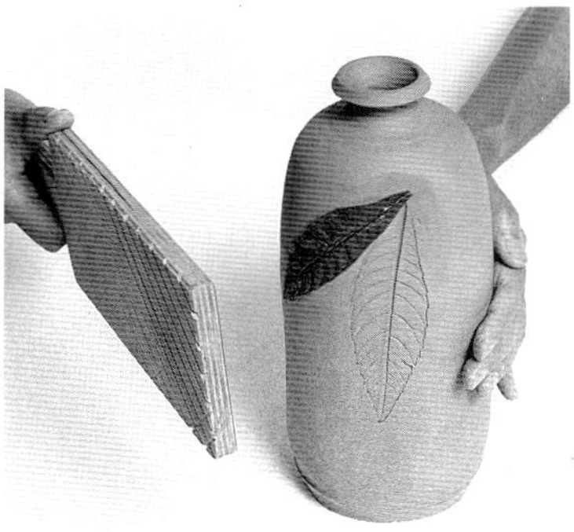

73. Leaf paddling

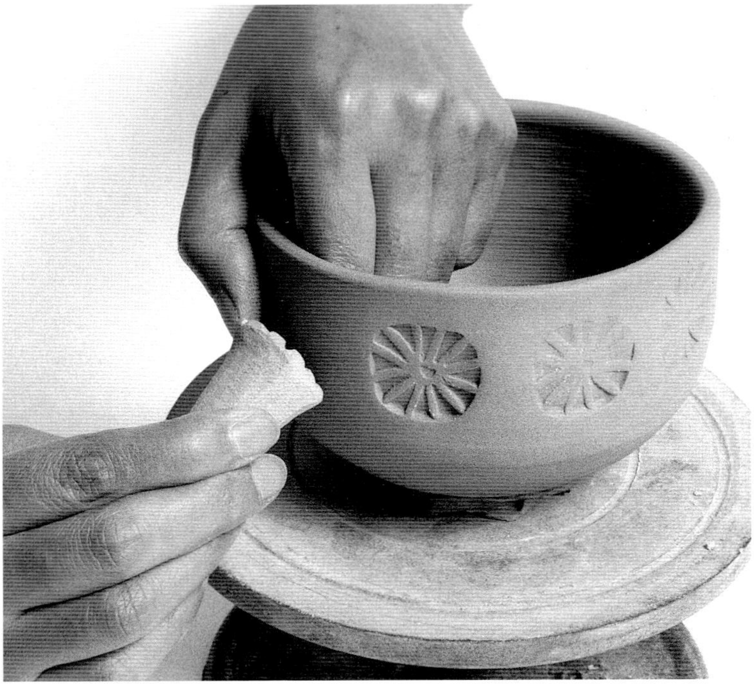

74. Stamping

Stamping

Stamping is another common type of impressed decoration. Stamps are usually made from biscuit-fired clay, but plaster and wood are used as well. The stamp is applied while supporting the wall from the opposite side (fig. 74). Roulettes, or rotating cylindrical stamps mounted on a handle, are also used.

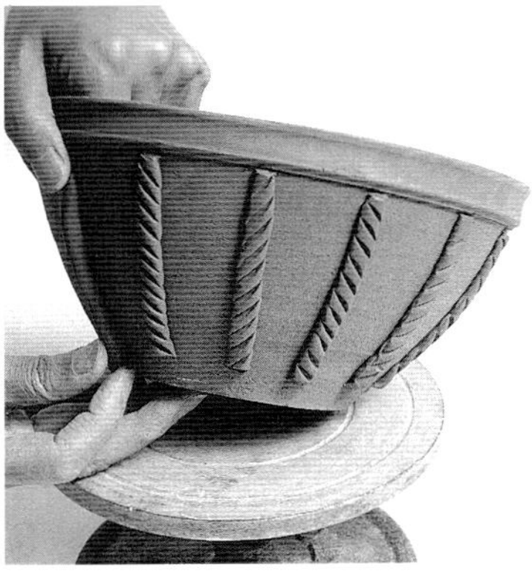

75. Coil appliqué

Appliqué

Since prehistoric times, various forms have been pressed or luted onto the exterior of the vessel. Modeling can occur before or after attachment.

Coil appliqué (fig. 75). The area to receive the coil is moistened, and the coil is applied with intermittent pressure from the index finger or with a tool. As so often happens, the technique, in this case the manner of pressing the coil, becomes a decoration in itself.

Stamped appliqué (fig. 76). Balls of clay are applied to the moistened vessel surface. A stamp is pressed into each of these. More elaborate additions can be sprigged (molded in intaglio molds) and then luted onto the vessel.

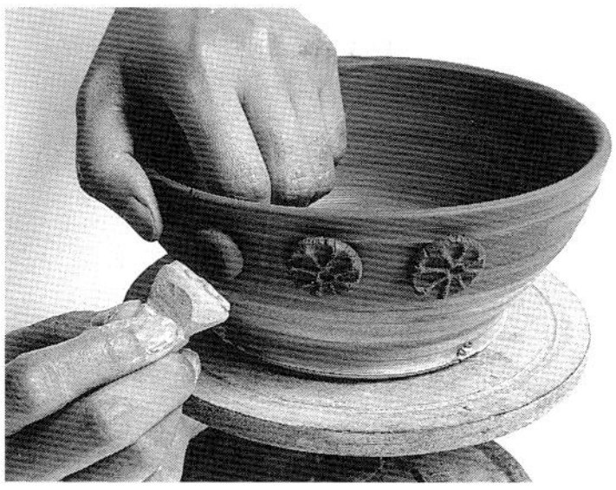

76. Stamped appliqué

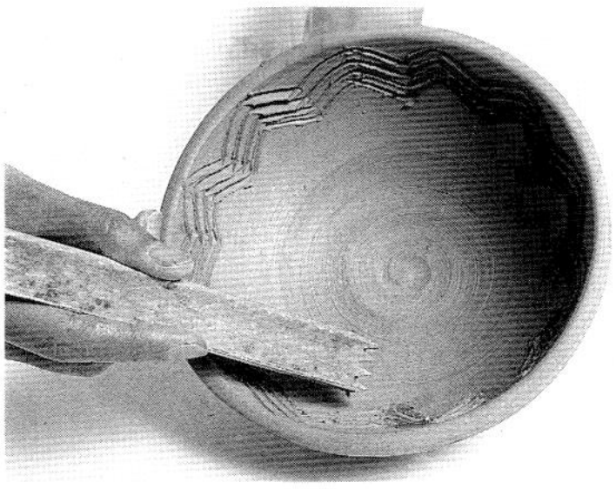

77. Combing

Incising

This technique is as ancient as Japanese pottery itself: the first potters used sharpened sticks and stones and the edges of shells.

Combing (fig. 77). The surface is incised with a wooden comb or a similar instrument. Such combed patterns form the abrasive surfaces on the Japanese kitchen mortar (*suribachi*).

Incising combined with forming (fig. 78). A cylinder is thrown on the wheel, and vertical lines are incised into the outside walls with one hand supporting the cylinder from inside (1). The wheel is started again, and the form is gradually widened. Since the outside has been combed, it is not possible to touch the exterior walls at this stage (2). The rim is finished, which also brings the pot back into center (3). Trimming is confined to the base, and a trimming chuck, which would damage the pattern, is not used.

78. Incising on the wheel

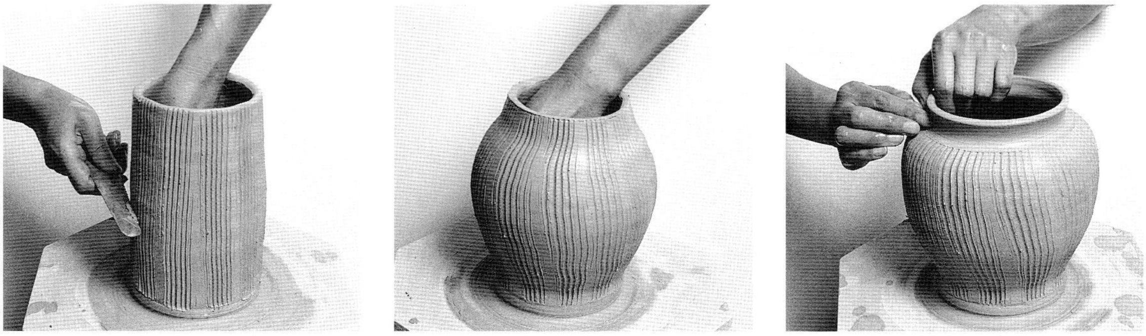

1 2 3

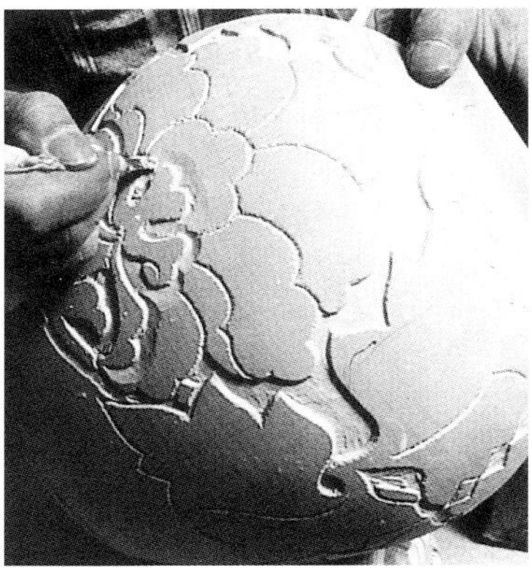

79. Relief effect

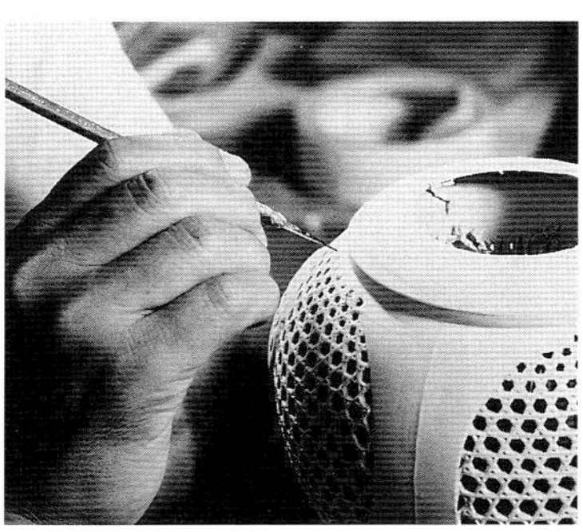

80. Openwork

Relief effect (fig. 79). The combination of linear incising with deeper, broader carving is called "floating incision" (*ukibori*). In this technique, the carving is deliberately variable in depth. This is especially effective with celadon glaze. The glaze pools and darkens in the deeper channels, and the thinly glazed areas seem to emerge in relief, or "float" on the surface of the vessel.

Openwork

Openwork, the perforation of the vessel wall for both functional and decorative purposes, has been used in Japan since prehistoric times. It appears consistently from the seventeenth century, first in Bizen and then in the Kyoto and Arita kiln centers.

The key to making good openwork is determining the appropriate time to cut into the wall. At the start of cutting porcelain vessels should be about 70 percent dry, and earthenware or stoneware should be about 50 percent dry. One effective cutting tool is a sharpened umbrella spoke fitted into a pencil-sized wooden handle.

After sketching out a composition, the potter opens holes in the vessel wall (fig. 80). The edges can be finished with a very sharp knife, although some professionals use watch springs as cutting tools.

A variation of openwork pioneered in China and now used in Japan is glazed openwork. This technique, used mostly with porcelain, features small holes, between 3 and 5 square millimeters in surface area. After the holes are perforated and finished, the vessel is bisque fired. The openings are then filled with a high-feldspar glaze, one that is translucent but not fluid. A formula of 85 percent feldspar and between 5 to 15 percent transparent glaze is the standard. The glaze suspension is thick, with a gum or seaweed syrup binder added. This glaze is brushed into the holes. The entire vessel is then covered with a standard transparent glaze and fired.

Burnishing

Burnishing was an important technique for prehistoric potters. Since it is a simple way to seal low-fired wares effectively, it was probably first used more out of necessity than out of any decorative impulse. A more conscious aesthetic is at work much later in Japan's red burnished tea wares *(shūdei)*. These wares were manufactured at Tokoname (Aichi Prefecture) and Yokkaichi (Mie Prefecture) for use with the steeped green tea *(sencha)* that became a popular beverage in the nineteenth century. The distinctive finish depends on a fine, fusible red clay. Now that the original deposits have been depleted, many potters add about 5 percent iron oxide and up to 15 percent feldspar to a finely levigated stoneware body.

Shūdei wares are polished when they are about 70 percent dry. One of the traditional polishing tools was a camellia leaf, also used by papermakers to put fine finishes on certain papers (fig. 81). Potters nowadays use a wide variety of materials, including nylon hose, steel, and polyethylene to polish the wares. Sometimes a brush is used to give a rougher texture (fig. 82).

No glaze is applied, and the wares are fired to about 1200 degrees centigrade in an oxidizing atmosphere. After firing, the wares are sometimes finished with very fine sandpaper.

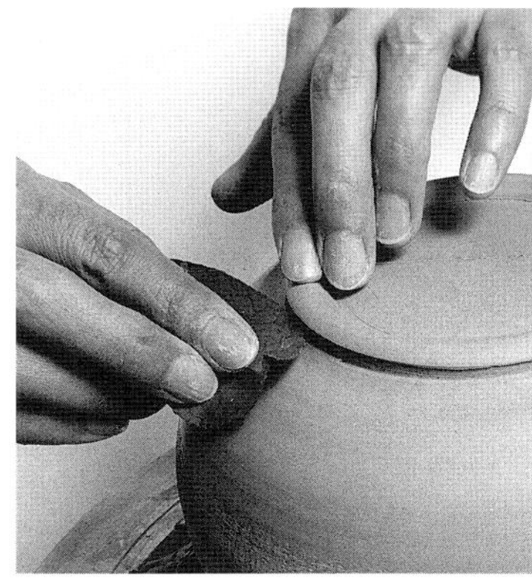

81. Burnishing with leaf

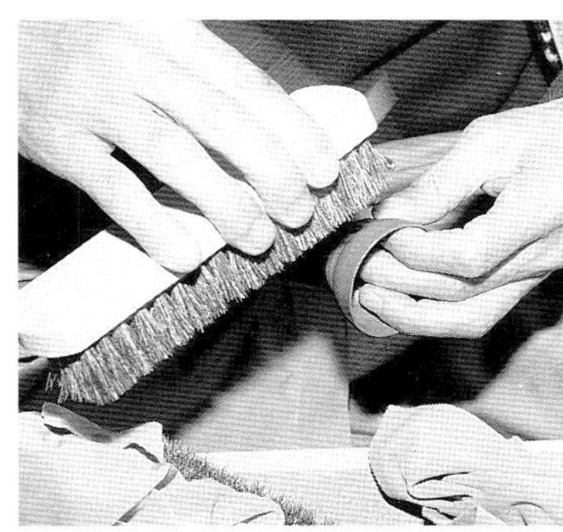

82. Brushed texture with burnishing

SLIP (ENGOBE)

The demand for white ceramics has been incessant throughout history. Yet pure white clays are scarce in comparison to impure, darkly colored ones, so potters found it convenient to liquify the former into a slip or engobe to coat the latter. In Japan, potters began to use fusible slips from about the fifteenth century, probably in an attempt to reproduce the glazes seen on foreign ceramics. White and red clay slips appeared as decorative elements on Shino and Oribe ceramics in the late sixteenth century, and the same period saw the introduction of slip techniques from Korea into Kyushu, and from Kyushu into Okinawa. By the early 1700s, Japanese potters were using white slip in three ways: as a covering for coarse, darkly colored clays; as a contrasting color field; and as a white pigment in underglaze painting. A fourth technique, that of using overall slip coatings to enhance the richness of colored glazes, was long a staple of Chinese production but was largely overlooked in Japan. The Japanese enjoyed the chromatic and textural contrasts between body, glaze, and slip, and the spontaneity and free play they found in the process of applying the material.

Formulating

The most important quality of a white slip, clearly enough, is that it fires white. Potters speak wistfully of the thin seams of white clay that occasionally accompanied deposits of darker clays in their localities. Barring access to this natural white slip clay, the most useful material is kaolin. Small kaolin deposits are still mined in Japan, but currently large quantities are imported from many nations, including Korea, China, New Zealand, and the United States.

Unfortunately, kaolin alone only rarely makes a good slip. Usually it requires additives: refractory stones such as silica or pyrophyllite to combat excessive shrinkage and maintain whiteness; fluxes such as feldspar or limestone to fuse the slip to the body; and plastic clays to facilitate adhesion in the unfired state or to help adjust the slip to body shrinkage. There is also the matter of timing. Slip may be applied when the pot is leather hard or dry hard or after bisque firing. At each stage, there is less body shrinkage. Accordingly, nonplastic material is increased and clay decreased and the clay portion may also require partial or full calcining. The following table suggests a basic approach to slip application in relationship to time of application.

	Leather-hard body	Dry-hard body	Biscuit
clay (kaolin and/or white plastic clay)	70 % (raw)	40 % (part calcined)	40 % (all calcined)
stone (silica or pyrophyllite)	20 %	40 %	40 %
flux (feldspar or limestone)	10 %	20 %	20 %

Lean clay bodies without much grog or sand will take a slip with a higher percentage of clay, while coarse clay bodies will need a slip with more refractory stone.

Japanese potters also make frequent use of *tōseki,* usually from Amakusa, in their slips. A natural mixture of quartz, fine potash mica, and feldspar, *tōseki* behaves both as a clay and as a flux. When ground well, it may develop a degree of plasticity and may be used as a principal slip ingredient with 10 to 20 percent additions of a plastic clay.

In Japan, colored slips were used before white ones. Medieval potters, for example, knew that fusible red clay slips would seal stonewares fired in the cool spots in their kilns. An understanding of the decorative potential probably derived from these utilitarian applications. Two examples of traditional colored slips are *ōdo,* a yellow ochre used on Red Raku ware, and *oni ita,* a ferro-manganic pigment used on Red and Gray Shino ware. These materials can often be found as thin precipitate layers between deposits of lighter-colored clays and soils. Mineral oxides such as cobalt and manganese may also be used to color slips, but traditional potters avoided such mixtures, possibly because those colorants affected the fusibility of the slip.

Off-white slips like those seen on folkwares from Okinawa and Kyushu are made by adding increments of an ordinary ball clay to kaolin and sometimes a small percentage of red clay (ochre) is added to a kaolinic white slip.

An interesting textural variant is called "sand dropping." Fine sand is sprinkled onto a slip application when it is still wet, and the excess is brushed off after drying, resulting in a granular texture.

Mixing

Before mixing slip ingredients, ball clays or any other plastic materials are dissolved separately. If the ingredients are still lumpy or grainy after immersion in water, they are ball milled. Otherwise the suspension is decanted from one bucket to another, and the remaining coarse material is crushed with a mortar and pestle. If the raw materials are of the processed type, a few hours on the ball mill will suffice. If the slip is ground too fine, it tends to scale off after application. The entire solution is filtered through a 100-mesh sieve. Before the days of wire mesh, potters used horsehair screens for clays and slips, and silk screens for glazes. When colorants are

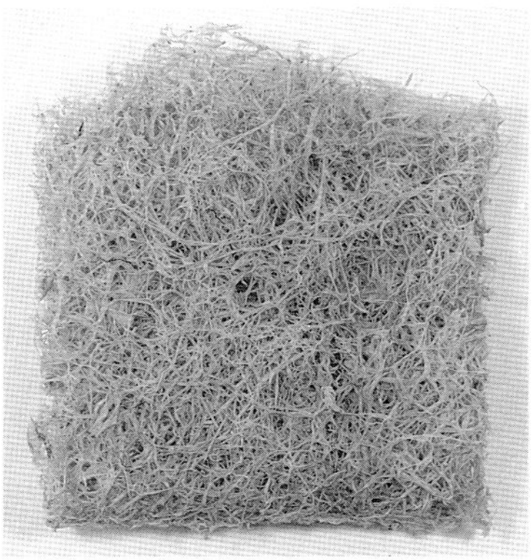

83. Sheet of dried funori

added to the slip, they are ground separately, as finely as possible, and then milled with the other materials.

Slips perform better if a syrup of *funori* seaweed is added (fig. 83). Its Western equivalent is Irish moss (gum tragacanth). CMC, or synthetic methylcelluose, is now used in Japan as well. *Funori* acts as a suspension agent and adhesive. It also has a slight deflocculating effect, reducing the amount of water necessary to make the slip fluid. The seaweed itself is sold dried, in thin mats about 40 centimeters square. Half of one of these mats yields 2 to 3 liters of *funori* syrup. Potters using smaller amounts mix a 10-centimeter-square mat of the seaweed into about 500 milliliters of water.

The syrup is made by placing enough seaweed to fill the pot into hot, but not boiling, water and simmering it until the solution is gelatinous. The solid remains are removed, the remaining solution is diluted until it pours easily, and then it is bottled and sealed. It has a powerful aroma. The more it is aged, the better it works. About 250 milliliters of the syrup suffices for a 20-liter batch of slip. After stirring in the *funori* syrup, the potter allows the slip to settle for a few days. Veterans maintain that when subject to long settling with *funori*, even a bad slip will surpass the good formulas.

Application

Testing indicates the right time for slipping raw ware: if the pot slumps and breaks apart when coated, the body is too moist, and if cracks appear when the rim is coated, it is too dry. Thickness and adhesion are tested on a throwaway pot. The slipped surface is scraped with a fingernail to gauge thickness.

Dipping (fig. 84) is the simplest and most widely used method of slip application for the outside of vertical forms and hollow ware. (When the interior of the vessel is to be slipped, that is done first, usually with the ladling method. The interior coat is allowed to dry so that the pot is rigid and the slip coat is no longer sticky to the touch.) To dip the exterior, the potter grasps the base of the pot, inverts it, and immerses it in the slip, all the time keeping it level (1). The pot is withdrawn slowly (2). When the foot is to be covered, the fingers are spread into the interior of the pot to hold it, and the pot

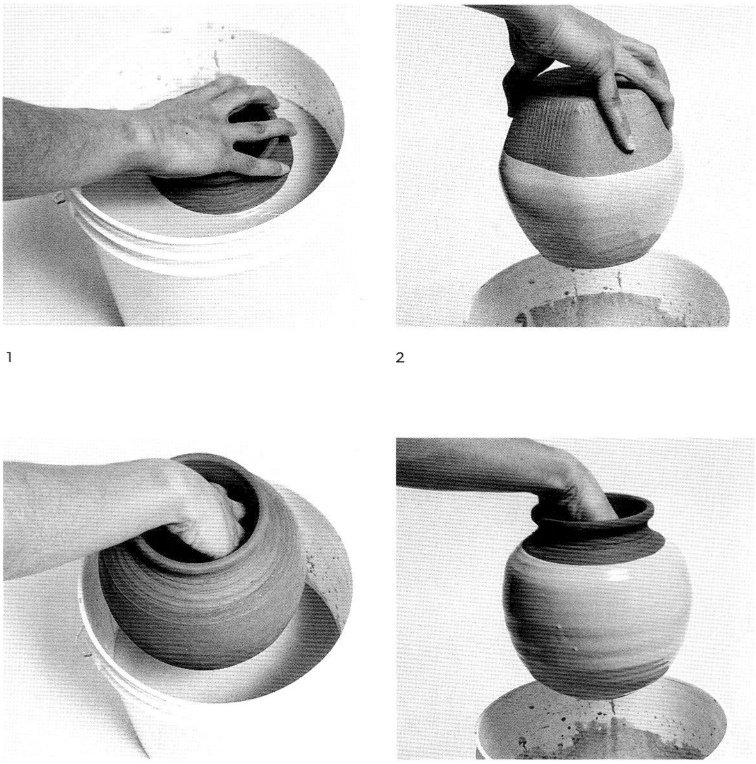

1 2 3 4

84. Dipping

is immersed foot first (3). When the desired coverage is attained, the pot is lifted out (4). The slip on the foot of the pot is lightly wiped off, and the pot is set aside to dry.

Ladling (fig. 85) is used to cover any surface that cannot be easily dipped into a bucket of slip. It is also used for decorative effects. The pot is tilted over the slip bucket. Slip is ladled onto the surface (1),

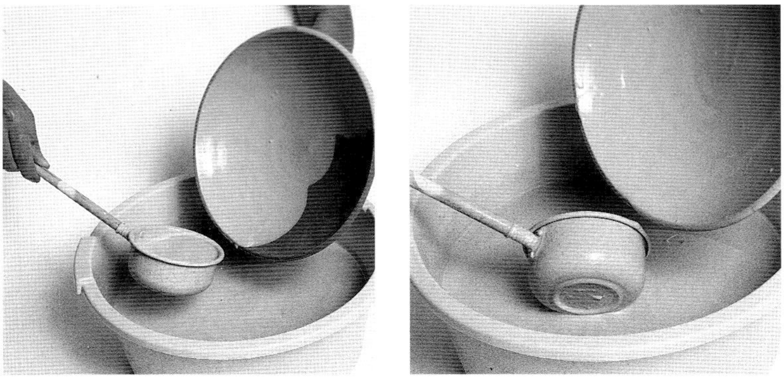

1 2

85. Ladling

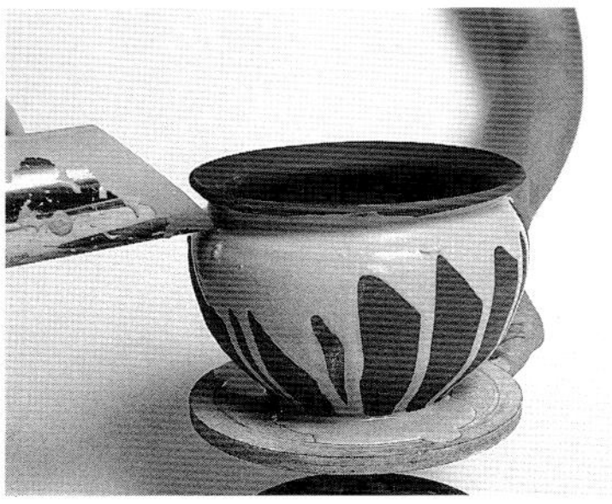

86. Pour patterning

which is coated by rotating the pot and letting the excess slip fall back into the bucket. The final drops that gather at the lowest edge of the rim are distributed by rotating that edge against something, such as the ladle (2).

The procedures for slipping a small bowl or cup, and for ladling slip onto the underside or exterior of a cumbersome vessel are identical to those of glazing (see pp. 133–37).

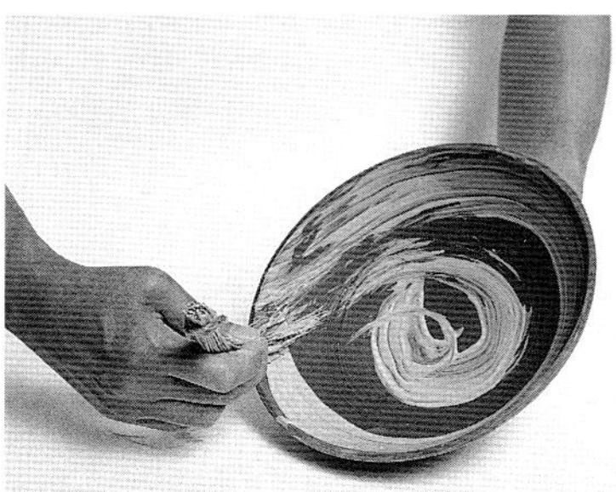

87. Hakeme

Pour patterning (fig. 86) is frequently seen on jars from rural kilns such as Tamba, where it was employed from the nineteenth century. The method can be used for glaze as well as slip. The pot is placed on a turntable or wheel and turned gently with the left hand. Slip is applied to the area just below the rim in a series of pours, controlled by quick tilts of the ladle cup (1). The Japanese ladle is useful here for its angled rim, which acts as a spout. Decorating a great number of pots in succession is the key to becoming adept at this technique.

Slip brushing (*hakeme*; fig. 87) originated not so much as a decoration but rather as a whitener for the dark clays used in Korea in the fifteenth and sixteenth centuries. Japanese connoisseurs prized the contrast between body and slip and even more so the uncontrived mark of the reedy brush. The best results are achieved when the clay body is high in iron (about 4 percent). The brush itself is little more than a bunch of dried straw, weeds, or twigs. Bristles from an old broom work fine.

A slip slightly thicker than that used for ladling or dipping is prepared. The surface of the pot is painted in a bold motion with a fully charged brush. Interesting effects are achieved on large pots by brushing first clockwise and then counterclockwise. Like pour patterning, *hakeme* gains life when decorating large numbers of pots. Such repetition allows an easy, uncontrived rhythm to develop.

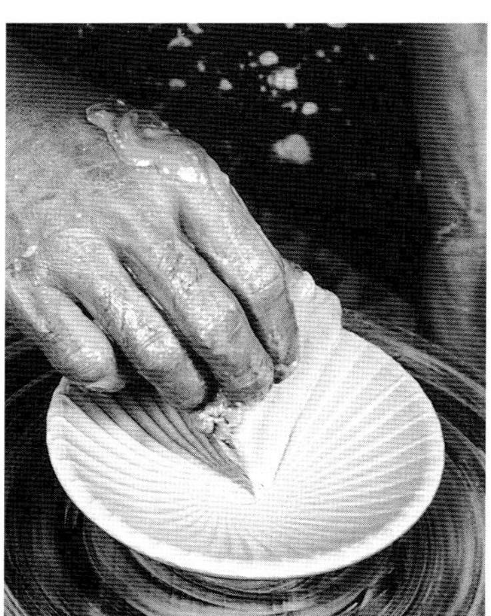

88. Slip patting

Slip patting (fig. 88) has been practiced since the nineteenth century in the workshops of Koishiwara and Onta in Kyushu. An iron-rich clay body brings out the richest contrast. A thick slip, similar to the one used for *hakeme*, is preferable. Broad, flat, chisel-type brushes are commonly used.

A small amount of slip is ladled onto a leather-hard pot. The turntable or potter's wheel is turned slowly. The brush, also charged with slip, is applied to the radius until the surface is evenly covered. The pattern is then made by striking the surface with quick, repetitive pats of the brush. This leaves tracks of thick slip interspersed with more lightly slipped areas. After firing, the latter will reveal the dark clay body. The method may be used on horizontal and vertical surfaces. Interesting variations may be obtained by combining different brush widths.

In a related technique, wet slip is combed with a tool or with the fingers (see pl. 23).

Slip trailing (fig. 89). Beginning in the late nineteenth century, makers of saké bottles inscribed the name of brewers or shop owners in a trailed slip (see pl. 24). This method, frequently but not exclusively associated with the Tamba kilns (at Agano there is a similar technique called "noodle trailing"), is carried out on dry-hard or biscuit ware. The applicator is a spouted bamboo cylinder about 25 centimeters long and 7 centimeters wide, although it can also be made from modern materials like plastic pipe. Since slip trails tend to flake off in drying and firing, extra *funori* is added to the slip, and the proportions of the batch are altered to assure that the slip will not scale off the body—about 50 percent clay, 20 percent pyrophyllite, and 25 percent feldspar or nephyline syenite is a good recipe. Some potters now add about 1 percent sodium silicate to deflocculate the slip, which increases fluidity without making the slip watery.

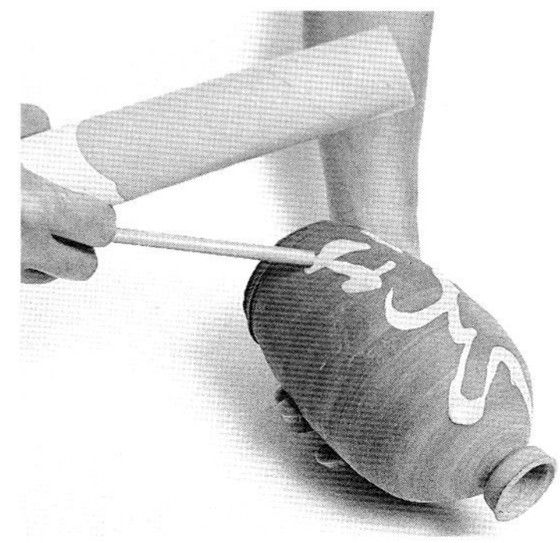

1

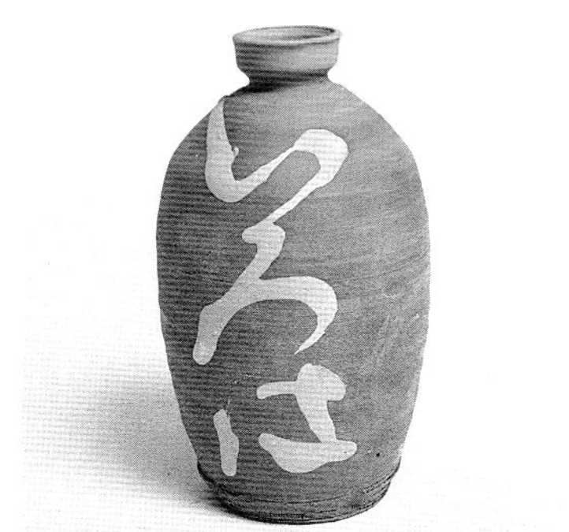

2

89. Slip trailing

The cylinder is filled about two-thirds and tilted so that a stream of slip issues from the spout. The angle at which the cylinder is tilted will determine the thickness of the line (1). The most control is achieved when the tip of the spout lightly touches the pot surface while decorating. The finished pot can be seen in (2).

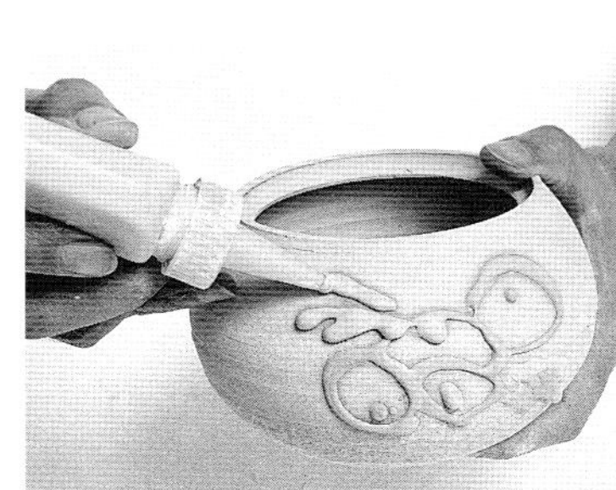

90. Thread extrusion

Thread extrusion (fig. 90) is the sophisticated urban equivalent of slip trailing. It was adopted from Kyoto textile dyers, who used trailed paste resist, in the seventeenth century. It is frequently used to produce cloisonné ware (called Kōchi after Cochin China, where the style is thought to have originated), a brightly colored lead-glazed ware with

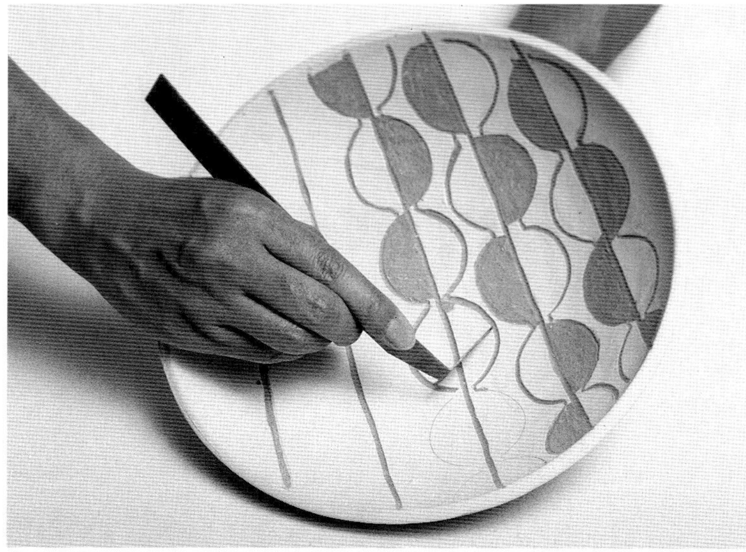

91. Sgraffito

extruded (or sometimes molded) thread designs in relief. The traditional extruder consists of a bag fashioned from paper that has been waterproofed with persimmon tannin and fitted with a brass nozzle. A similar effect can be obtained using a spouted plastic bottle.

In the Kyoto version of thread extrusion, the pots are first fired to a hard biscuit, about 1200 degrees centigrade. The slip is thick, pasty, of even consistency, and high in stone and flux. A mixure of 40 percent clay, 30 percent Cornwall stone, 20 percent pyrophyllite, and 10 percent feldspar works well. To prevent excess shrinkage of the slip, which causes scaling, the water content is kept at a minimum: this is achieved by adding *funori* syrup or sodium silicate, perhaps in combination. The slip is very carefully screened to avoid clogging of the nozzle.

After determining the composition, threads are squeezed out onto the surface of the pot. For intricate designs, the outlines are drawn first with a pencil. The cloisonné effect is obtained after a second bisque, when overglaze enamels are applied using the raised lines (cloisons) as borders. In many rural kilns of the premodern era, a more vigorous thread extrusion technique was used to decorate dark-bodied stoneware with simple motifs of flowers and grasses.

Sgraffito technique (fig. 91), first used to great effect in the Cizhou ware of Song-dynasty China, is produced by applying a slip that contrasts with the clay body or with another slipped surface and carving through that application to create a pattern. In Japan, it first appeared at the end of the sixteenth century in Shino ware.

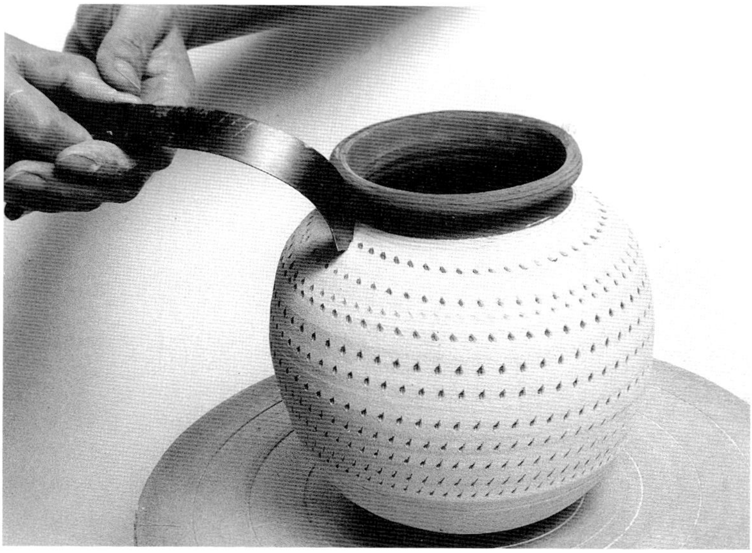

92. Chatter marking

The vessel is slipped and dried to a leather-hard state. A pattern is sketched, and then unwanted slip areas are trimmed away with shallow strokes of the trimming knife. Either the motifs or the ground may be trimmed, although it is usually the latter. This technique also works well with a single-application slip glaze carved through to a contrasting body.

Chatter marking (fig. 92) was first used in China around the twelfth century. In Japan it appears in Okinawa and Kyushu folk kilns from the eighteenth century. It is made by applying a springy metal tool to the surface of a slipped pot. The tool bounces and creates a rhythmic pattern of nicks in the slip, exposing the contrasting clay underneath.

As the pot revolves on the wheel, the blade is applied to the slipped surface. It takes some testing and experience to determine the correct state of dryness, the correct speed of the wheel, and the desired chatter pattern. If the slip coat is too wet, the blade won't skip. If too dry, the surface becomes impenetrable.

Slip inlay (fig. 93), the technique of inlaying slip into recessed areas, was transmitted from Korea to Japan in the seventeenth century. A dark clay body, with at least 4 percent iron content, is preferable and the slip suspension is kept thick.

Patterns are stamped into the surface with one hand reinforcing the opposite side of the pot (1). The impressed areas are painted over with a thick white slip (2). The pot is dried to a leather hardness, and then the slip is scraped away with a blade (3). The timing of this last

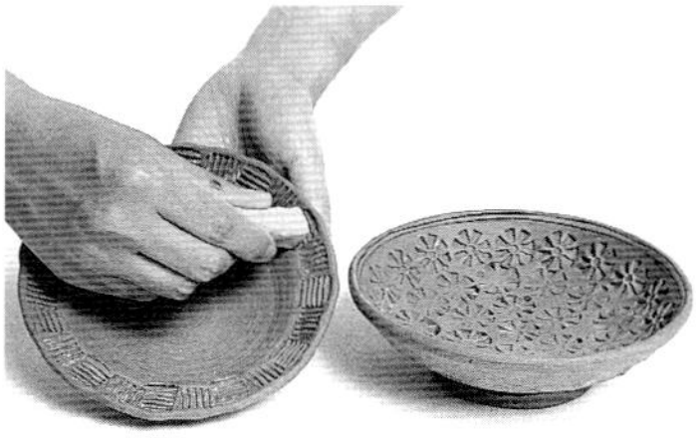

1

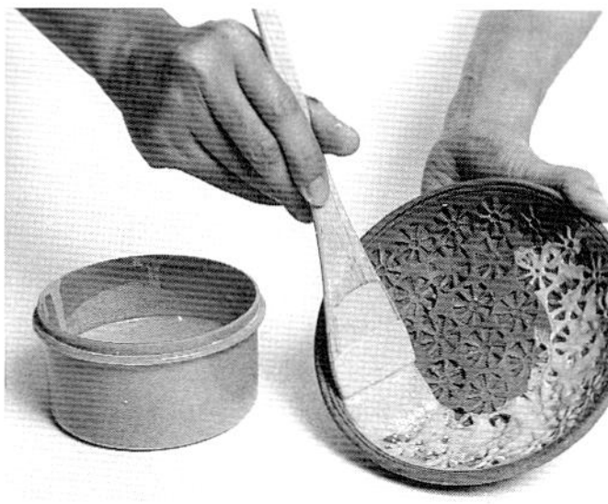

2

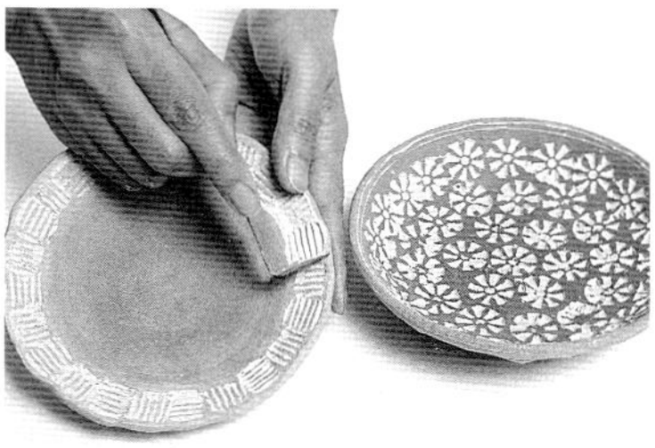

3

93. Slip inlay

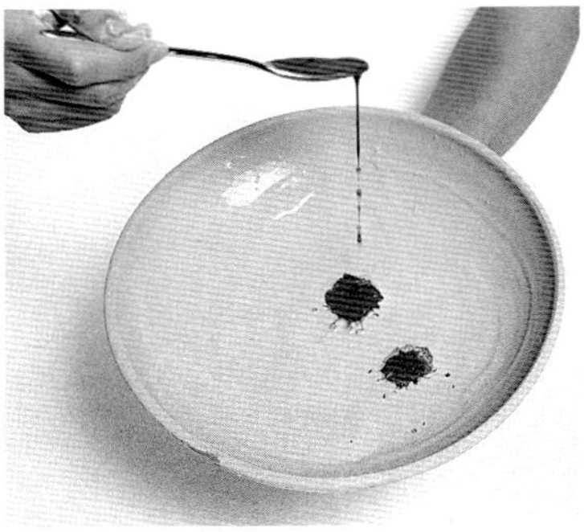

1

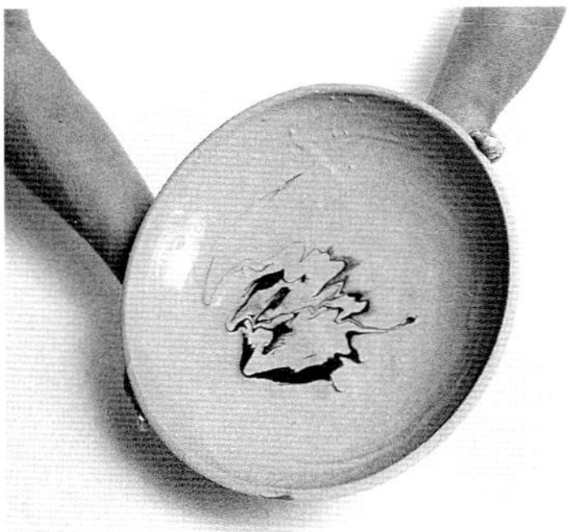

2

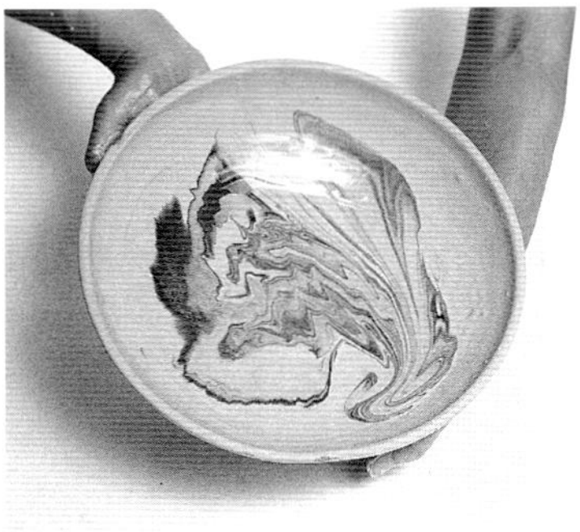

3

94. Slip marbling

step is important: too early and the trimmings will stick to the body and slip; too late, and the pattern contours will not be crisp.

Slip marbling (fig. 94), appearing first in eighth-century Chinese earthenwares, became a staple of the Tamba kilns in the nineteenth century.

White slip is applied to a leather-hard pot. The coating is a little thicker than usual. A small amount of colored slip is immediately poured on top of the white (1). Then the pot is shaken so that the colored slip glides over the white, creating a marbled pattern (2). Since the raw clay vessel will quickly soften, the handling is very brief. The finished pot is as shown in (3).

Another variant of slip marbling, really an old English technique, is feathering. Slip lines of a contrasting color are applied over a wet slip base with a syringe. Then a pointed instrument is used to drag lines perpendicular to the trailed ones. This produces a featherlike pattern.

UNDERGLAZE PAINTING

Underglaze painting is a rather late development in the Japanese tradition. This is partly due to the earlier development of monochrome glazes, which firmly established color and texture as primary points of interest. There is also a technical factor: transparent glazes, which are necessary to show underglaze painting, are harder to produce than monochrome glazes. In Japan, feldspathic stoneware glazes, glazes that are sufficiently transparent to reveal the painting underneath on one hand and sufficiently refractory to hold the pigments on the other, were produced from the last quarter of the sixteenth century. These early pigments were types of iron oxide. Painting in underglaze cobalt (blue-and-white) was developed soon after, probably in the second decade of the seventeenth century.

Underglaze Iron

Iron pigments are the easiest to find and the easiest to process, and until recently, they were most frequently used, especially on stoneware and earthenware. Traditionally potters used either naturally occurring iron-bearing clays and stones or iron-oxide scale from a blacksmith's forge. Japanese potters use six or seven major varieties of iron sources. *Oni ita* ("devil board"), used in Shino and Oribe wares of the Mino region, is an impure clay containing about 40 percent iron and found in thin layers between other materials. *Mizutare* ("water gravy") is a hydrated iron precipitate, well over 50 percent iron, that oozes out of larger clay masses and can be seen welling up in rice paddies and bogs in the spring. It is a favorite of Kyoto and Shigaraki potters. *Ōdo* (yellow ochre) is a red or ochre-colored clay that averages about 10 percent iron. There are many other materials, ranging from 6 percent to 90 percent iron content. Some of these mixtures impart just the right balance of refractoriness, color, and ease of application for the decorator. Naturally the ones that form a suspension in water are preferred. When used under a glaze, most of these materials fire to a shade of black or brown.

A blend of iron oxide and ochre in equal parts makes a good underglaze iron pigment. Diluting the batch with pyrophyllite or calcined kaolin will make the color paler. Mixing iron scale with an equal amount of natural cobalt creates a black pigment. In all cases thorough grinding is necessary.

The metallic "rust" flashing seen in much underglaze-iron painting in Japanese stonewares results not so much from the quality of the pigment as from the quality of the glaze, especially the amount of calcium it contains and the thinness of its application. At high temperatures, a calcium-rich glaze (for example, a wood-ash glaze) attacks and dissolves the iron oxide in the pigment. Since the glaze can only hold a certain amount of iron in solution as it cools (about 3 to 4.5 percent in normal cooling conditions), surplus iron is precipitated in the form of iron crystals, which provide the rust effect.

Underglaze Cobalt

This technique was transmitted from China and Korea to Japan in the early decades of the seventeenth century. In China, it has been a great favorite since the 1300s. Although underglaze cobalt has been used on stonewares and earthenwares in Japan, it is generally considered to be most effective on porcelain. Accordingly, the history of cobalt decoration and porcelain manufacture are intertwined. Once the Japanese had mastered the technique and adapted it to domestic taste, it quickly eclipsed underglaze iron-decorated stoneware in popularity. By the end of the seventeenth century, blue-and-white decorated porcelains, especially those from the Arita kilns in Kyushu, dominated the upper- and middle-class markets, and iron oxide decoration was relegated to folkware.

In contrast to the relatively pure cobalt oxide used in the West, the Japanese traditionally paint in natural cobalt (asbolite, or *gosu*). This contains various oxide impurities in addition to the cobalt. The composition averages about 40 percent silica, 25 percent alumina, 25 percent manganese, 5 percent iron, and 5 percent cobalt. The mixture tends to soften the blue color of the cobalt and under a glaze produces the quiet purplish or grayish blues so prized by porcelain connoisseurs. Although most *gosu* was imported from China, small amounts were found in Japan, mostly in Seto.

Most potters now purchase synthetic *gosu*, but it is not difficult to mix a batch in the following proportions: 20 percent cobalt oxide, 10 percent manganese dioxide, 10 percent iron oxide, 10 percent nickel, and 50 percent kaolin. Everything except the kaolin is mixed with water and ground in a mortar. The mixture is dried, fired to about 1200 degrees centigrade in a crucible, and ground again. The kaolin is added and the mixture is given a final grinding. For a strong blue, the percentages of cobalt and manganese are raised. For a subdued color, the percentage of iron is raised.

A quiet blue can be made by adding a small increment of cobalt oxide (about 2 to 5 percent) to an impure clay-like yellow ochre and grinding it well.

Complex painting schemes in underglaze blue often call for variable color saturation. The traditional thinner is clay, but this causes the pigment to flake off after application. Some potters use porcelain stone, but this lacks refractoriness and causes a diffuse, bleeding effect, of a type that can be seen on Vietnamese ceramics from the fifteenth and sixteenth centuries. Most Japanese workshops now use pyrophyllite, in increments of up to 30 percent. When pigments bubble in use, they probably need a prior calcining, and possibly dilution with clay or pyrophyllite.

Traditionally, cobalt pigments were subject to a great deal of grinding. One can still find a few old hands who object to pigments ground in ball mills, for machine-milled round particles lack the plasticity of the irregular shaped ones that result from hand-processing. Veterans maintain that it is more difficult to paint with machine-milled pigments. Most underglaze-cobalt painters suspend their pigments not in plain water but in a solution of thick left-over green tea. About 10 grams pigment to 180 milliliters of tea is a good starting point. The tea base also adds a kind of elasticity to the pigment and helps it stick to the pot. Predictably, some contemporary potters have substituted instant coffee for the tea.

Brushes

Most Western potters are by now familiar with the physical and aesthetic properties of the East Asian brush. Although some potters use painting or calligraphy brushes, they lack the strength and absorbency needed for extensive painting with ceramic pigments, and certain specialized types of brushes are used by decorators. The outlines and linear details in underglaze cobalt painting are drawn with a brush called a *mensō fude,* which has a long, thin hair cluster. The wash areas between dark outlines are painted with a fat brush called a *dami fude.* The handles of these brushes hold a lot of pigment, which is gently squeezed out in the course of application.

If these decorator's brushes are not available, good brushes can be made from animal hair, or from store-bought Asian brushes. The best time to gather animal hair is early summer, when it is rigid. Preferred sources include the dog (tip of the tail, back of the neck), mouse (tufts under the legs), deer, cat (back), and fox (back and shoulder). Whether the hair comes from animals or disassembling Asian painting and calligraphy brushes, the brush hairs are arrayed tapered ends out with the longer hairs in the center. Then they are bound at the base with a thread or a strip of strong paper and inserted into a holder or handle. This handle need not be the traditional bamboo; the split end of a dowel works well.

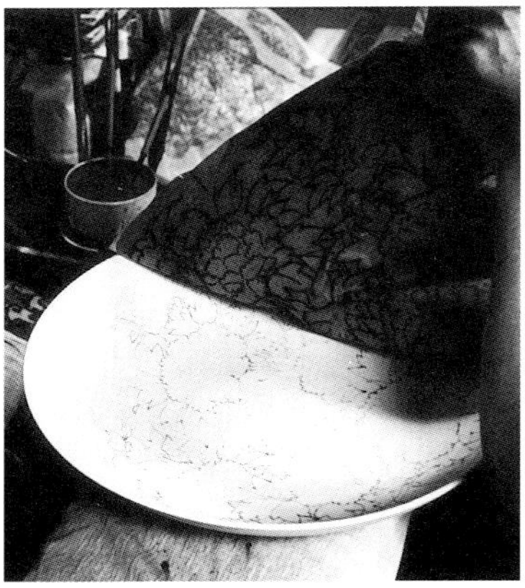

95. Charcoal transfer

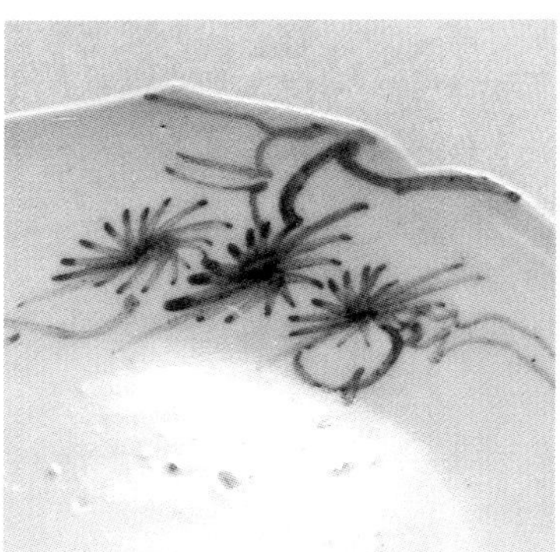

96. Linear brushwork

Application

Bisqueware is dusted off before it is painted. A handy duster can be made by tying several strips of cloth about 4 centimeters wide and 25 centimeters long to the end of a short handle.

For designs produced in limited quantities, compositions are sketched in a solution of powdered charcoal (a thick ink will stain the piece), and the painting is done over that. Potters making larger quantities of underglaze-blue-decorated porcelain traditionally have produced a stencil through which cartoons can be transferred to each piece. To make such a stencil, a piece of strong, long-fibered paper (traditionally, handmade mulberry paper) is cut to a shape that covers the design area of the pot. A master design is painted on one pot, and the paper is pasted over that with *funori* seaweed syrup. The *funori* syrup will make the paper transparent, revealing the cartoon underneath. The areas that cover the painting are carefully cut away. The paper, now a finished stencil, is removed and dried, and a dilute, fine charcoal solution is painted onto its inner surface. The stencil is then applied to a succession of pieces (fig. 95). The charcoal is transferred to the pot surface by rubbing the outer surface of the stencil with a soft object such as a leaf or cotton wad. About twenty cartoons can be transferred with every charge of charcoal. Finally, the areas left blank are painted.

Cobalt for painting is usually kept in a small porcelain mortar, and is stirred occasionally with a pestle. Bowls and flatware are decorated on small banding wheels, while hollow objects such as flower vases are sometimes inserted over a horizontal armature padded with cloth to protect the inside of the pot. Pots can also be held in the hand, with a thin cotton glove worn to minimize smudging. Decorators sometimes wedge a short, flat board between their elbows and lap and use that as a base for decorating small objects. For fine painting, the wrist rests against the pot or another close support. Simpler painting uses the whole arm. The potter's wheel or small banding wheel may be used to make horizontal bands. To make circles, the brush may be attached to a compass.

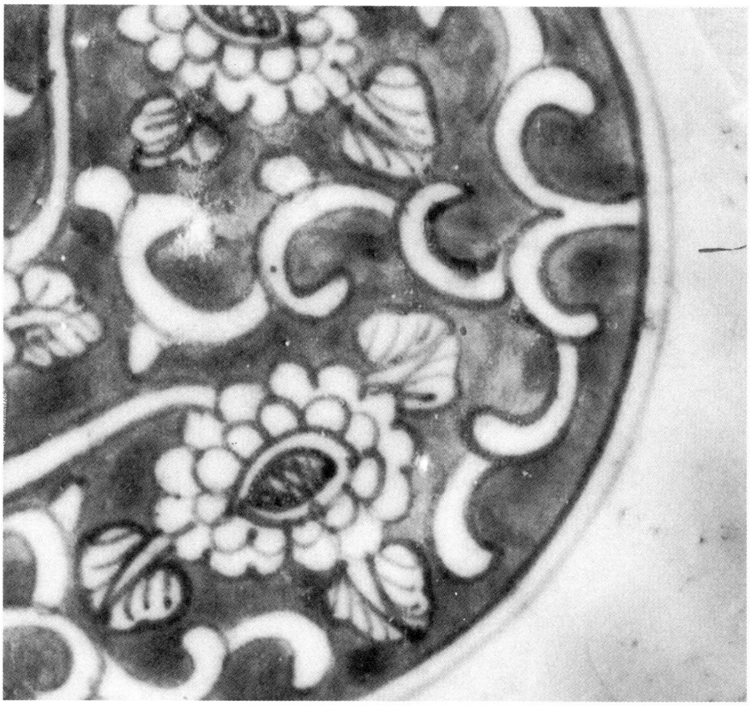

97. Outline and wash painting

The most common methods of underglaze blue decoration are as follows:

Linear brushwork (fig. 96). This the simplest and most spontaneous method, often seen on folk wares.

Outline and wash painting (fig. 97). The outlines are painted first. To keep the surface clean, the painting is begun in the center of the composition and expanded outwards. For the solid wash portions, some of the regular pigment is diluted, inserted in the hollow tube of the large *dami* brush, and applied. This wash is not really brushed on, but applied in a jotting motion. The brush hairs are periodically squeezed to supply more pigment to the tip.

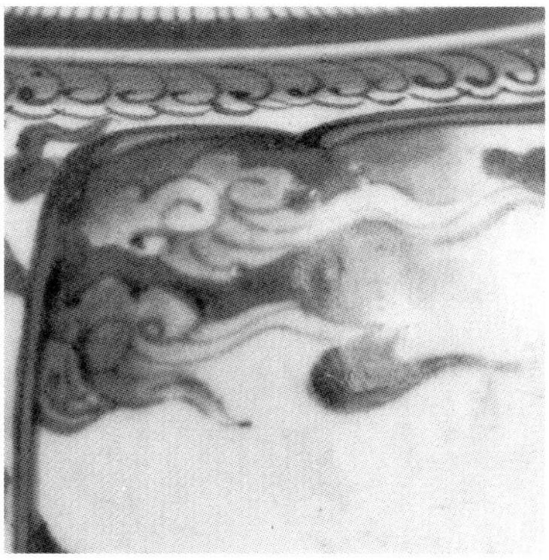

98. Graded washes

Graded washes (fig. 98). Ink painters can achieve gradation in a single stroke using an irregularly charged brush, but with ceramic pigments it is easier to combine separate dark and light strokes. The brush is usually held at an oblique angle. A dilute pigment is first applied to the areas that are to be light. A thicker pigment is applied alongside the light areas to create the darker shade. Gradation is also accomplished by painting first in regular cobalt pigment and then selectively diluting that with a brush charged only with water.

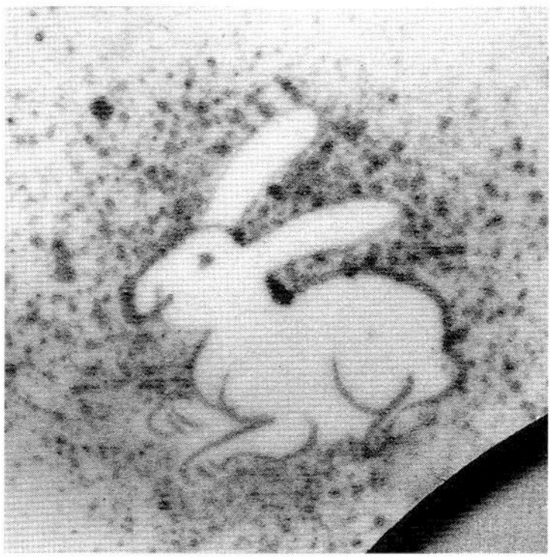

99. Spray effects

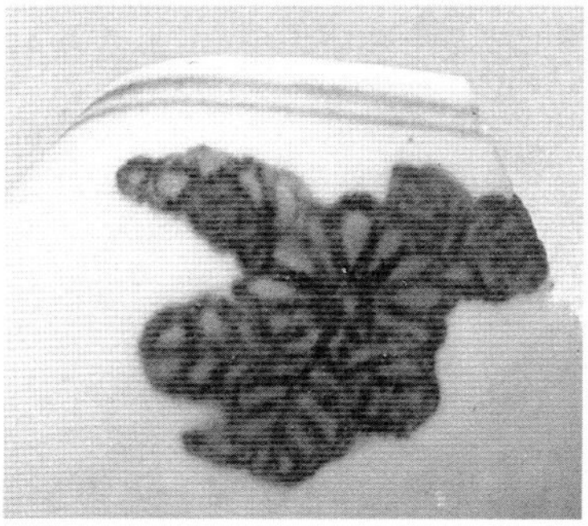

100. Stamping

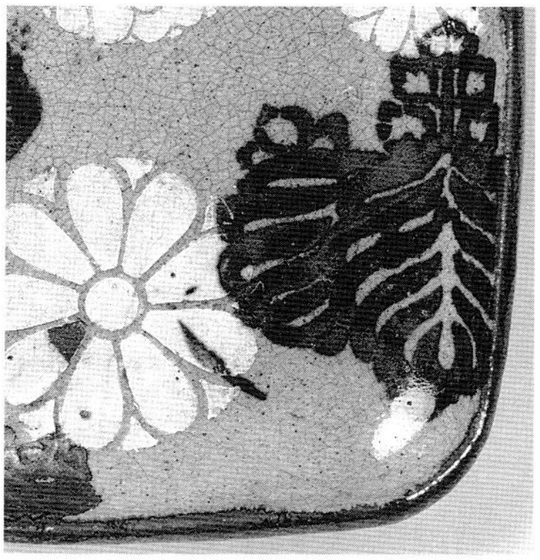

101. Stenciling

Spray effects (fig. 99). Atomizing was transmitted from China to the Arita porcelain kilns in Kyushu in the early seventeenth century. It was also used by Seto folk potters in the eighteenth and nineteenth centuries. Pigment—or as was occasionally the case in China, dry powder—is blown onto the vessel through a tube covered with coarse gauze. It can also be blown off or flung from a large brush tip. This technique is often combined with stenciling.

Stamping (fig. 100). Pigment is applied by means of stamps of the gum type. In the past some potters may have fashioned stamps from *konnyaku*, a thick, rubbery foodstuff made from the root of the devil's tongue plant.

Stenciling (fig. 101). Pigment is applied through or around stencils traditionally made from paper waterproofed with persimmon tannin. Today, many potters use synthetic varnish as waterproofing. The stencil is lightly glued to the pot—it is particularly important that the edges are held fast to the surface—and pigment is applied.

Etching (fig. 102). The pigmented surfaces are incised to reveal the body color underneath. This can be done at the raw or biscuit stage, but the quality of the line will vary considerably.

Wax resist does not have a long history in Japanese ceramics, but it has been very popular over the past few decades. Until recently potters heated a potentially dangerous mixture of paraffin and kerosene, but now they use liquid latex. If lines are incised through the wax, decoration can also be created inside the resist field. Unless the decoration is carried out on the raw ware, the piece has to be bisqued a second time to remove the wax. Otherwise the glaze will not cover the resisted area.

A related Arita technique is "ink resist"—*sumi hajiki* (fig. 103). Fine linear patterns are applied in ink prior to the bisque fire, and a cobalt wash is applied over that. In the bisque firing, the wash scales off the ink, leaving linear patterns. This is the best way to create thin, fluid lines in resist.

Underglaze blue works best with a "hard" glaze, one refractory enough to hold the contours of the pigment. Traditionally, potters used ash made from the *isu* tree (see p. 130–31), which was very refractory, but now limestone glaze is the common favorite. Lower-melting wood-ash glazes attack the contours and alter the colors of cobalt painting. Most blue-and-white kilns are fired in reduction.

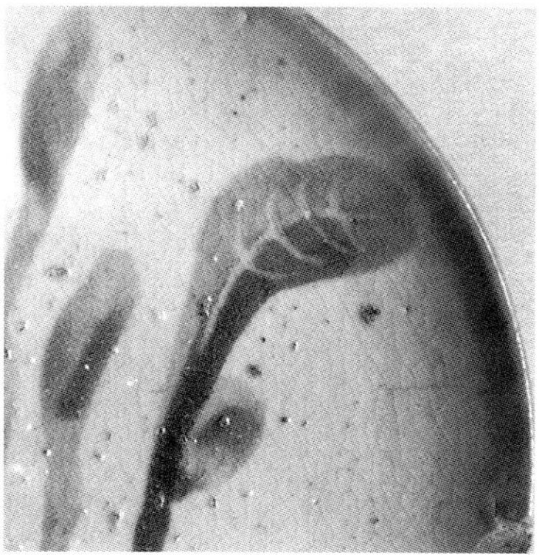

102. Etching

Underglaze Copper

This appears only sporadically in traditional Japanese ceramics, usually in its oxidized green rather than its reduced red form. Its first use occurs in the diffuse green accents in Yellow Seto ware of the late sixteenth century, then in the seventeenth century in Karatsu-lineage wares in Kyushu. In the nineteenth century it appears with more frequency in folk wares from many locales. Potters first obtained their copper from a naturally occurring copper sulphate, which was baked to drive off the sulphur. In more recent times, potters have oxidized copper from scraps of metal by immersing scrap metal or filings in salt water for about a week. The mixture is stirred several times a day. The powdery oxide that forms on the metal is rinsed well to remove the salt. To remove any further traces of salt, the powder is soaked in old tea and rinsed again. Decorators made a pigment by mixing about 10 percent of this copper powder into feldspar. This traditional recipe tends to blur a little and may be made stiffer by adding a refractory stone like pyrophyllite. Most potters now use barium in copper pigments and glazes, since this tends to enhance the color of copper in its reduced state. Copper is very unpredictable, and the accidental colors that can result—for example the dull liver color that occurs when copper is exposed to a middle flame—did not endear the technique to Japanese potters.

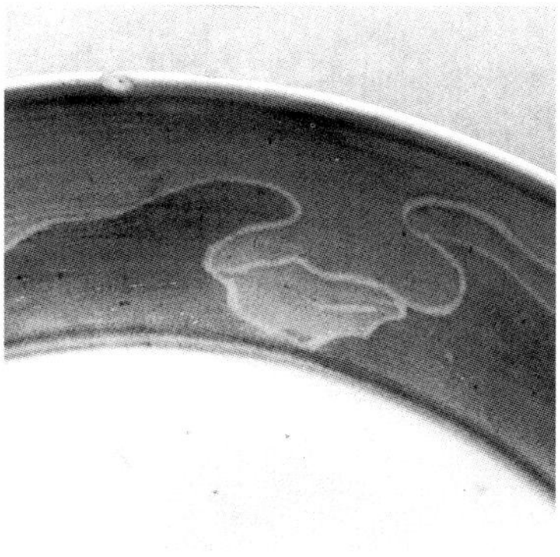

103. Ink resist

FIVE

Glazes

Historical records and heirloom pieces show that by the eighth century Japanese potters understood how to coat their vessels with a powdered glass which would waterproof the ware, decorate it, and keep it clean after firing. This early glaze, a mixture of lead oxide and siliceous rock powder, fired to a transparent gloss at a temperature of 750 degrees centigrade. The potters also discovered that small additions of mineral colorants such as copper or iron would yield green or brown glazes, respectively. These compositions, however, were of little use at high temperatures, which caused them to volatize. In the late eighth or early ninth century, Sue-ware potters noticed that fly ash from their wood fuel formed glossy areas on their stoneware pots and began to apply the ash deliberately. This first took place at Sanage, in present-day Aichi Prefecture. Judging from the surface quality and greenish color of this early glaze, it probably derived from a softwood used in the firing of the kiln. But ash alone was too fluid at high temperatures, so a refractory material was needed to balance it. Some four hundred years later, the potters at nearby Seto achieved this by adding an increment of buff clay body into the glaze: the formula consisted of about 10 percent body to 90 percent glaze. A century later, potters substituted a high-iron clay for the buff, and a brown-black glaze resulted.

The use of these ash-clay glazes persisted in central Japan, particularly in the Seto region, until the sixteenth century. Then new technical influences from Korea encouraged the use of feldspar as a principal glaze ingredient. When combined with hardwood ash, feldspar melted into a transparent glaze that permitted underglaze and overglaze decoration when used over light-colored clays or slips. Overglaze enamel technology, based on the use of a prefired glass called frit, was imported from China in the 1640s, first appearing in Arita and shortly afterward in Kyoto. From the late nineteenth century, high-production potters gradually abandoned wood ash in favor of limestone, which is more consistent in its chemical composition.

There are many ways of developing a stable of glazes. One can work from empirical formulas established in Western glaze chemistry, or one can try out batch recipes, of which there are thousands published. The best approach to Japanese glazes, however, is suggested by their history. The potters started making glazes with ash alone, then over the centuries variously added clay, ferrous stones, feldspar, and later industrially refined minerals and metallic oxides. Reviewing this process will provide a basic understanding of Japanese glazes as well as a basis for working with materials available at any locale.

[Note: All batch formulas or ratios are dry percentage weights. All firing temperatures, unless otherwise indicated, approximate Seger cones 8 through 10 (Orton 9–11). Firing schedules and atmospheres will be discussed in the next two chapters.]

ASH GLAZES

The traditional Japanese potter made glazes with local ashes, clays, and stones, either alone or in combination. Potters produced large amounts of ash in firing kilns. Hearths and kitchen stoves were rich sources of wood ash as well. Since rural potters doubled as farmers, ash from rice straw and rice chaff, customarily burned to refertilize the paddies, was also plentiful. In metropolitan centers, ash became available through a network of intercraft cooperation. Indigo dyers, for example, used the soluble part of hardwood ash to make lye for dyeing and passed on the solid-ash remains to potters. The Kyoto neighborhood I lived in was populated by two groups of artisans: potters and fan makers. The Japanese fan is mounted on thin bamboo staves, and in trimming these staves the craftspeople produced huge amounts of bamboo shavings. The potters obtained these shavings and burned them for a high silica ash. Ash can be obtained this way in many parts of the world. In Hawaii, I obtained ash from the local macadamia nut plantation, which burned the oily shells for fuel. In Houston, barbecue restaurants were happy to provide ash from the mesquite wood they used for cooking their foods.

Two main types of ash are useful to potters. One type is wood ash, which is typically high in calcium, and the other is straw ash (especially from rice, wheat, barley, and rye straws), which is typically high in silica. Within the wood-ash category, softwood ashes are typically higher in calcium than hardwood ashes. Even within a single species of tree, however, there will be considerable variation depending on the age of the tree, the time that it is harvested, and which part of it is used. Bark and branches are traditionally considered to yield the best ash: they contain about twice as much calcium as the trunk and leaves.

Ash is processed by first burning the raw material in a clean place—a bed of clean firebrick is ideal—or by collecting the ash from a dependable source. The ash is put in a large bucket of water and stirred. The carbon and scum that float to the surface are screened off, the liquid portion is decanted, and the heavy grains that remain on the bottom are discarded. Every day the liquid at the top is decanted and replaced by clean water, after which the mixture is stirred and left to settle again. Warm water speeds up the process. The slimy feel that comes from the ash solubles, which variously include carbonates, chlorides, and phosphates of sodium and potassium, gradually disappears, and the water becomes clear and tasteless. (There is a debate about leaving these solubles in the ash. They contribute toward interesting but unpredictable effects due to their penetration into the clay body.) To get a particularly fine ash, some potters first make the wood into charcoal and then burn it into ash. For potters with sufficient reserves, ash is stored before use, up to five years. It is said to reach its prime upon turning a crusty brownish yellow.

Pure Ash Glazes

A simple ash glaze may be obtained by dusting a pot with ash or dipping it in an ash suspension. A binder will probably be necessary. A thin wood ash coating, which is mainly calcium, will combine with the silica and alumina in the clay body to produce a thin glaze. If solubles are left in the ash, they will interact extensively with the body, bringing, for example, concentrations of iron to the surface. This is the cause of the thin, dark glossy area that forms at the glaze-body interface. If this simple ash glaze is applied thickly, surface tension caused by the high concentration of calcium often results in a beaded effect. These beaded areas in turn tend to crystallize, an effect known as devitrification. If there is extensive crystallization, it may diffuse reflection and make the glaze opaque. Pure ash glazes also tend to run.

Ash and Clay Glazes

To stabilize the fluid ash glaze, potters added the most refractory material at hand. For the premodern Japanese potter, this was usually clay. The result was a glassy but still fluid greenish glaze such as the type seen in Sanage and Seto wares (see figs. 9, 11). As mentioned previously, this can be recreated by adding about 10 percent low-iron clay to 90 percent softwood ash, especially pine. As potters experimented with proportions and clays, a whole range of effects was achieved. One enduring Japanese favorite is *irabo* glaze, first seen on sixteenth-century Korean stonewares but since modified by Japanese potters (fig. 104). It is achieved by combining about 50 percent wood ash with 50 percent of an iron-bearing clay such as ochre. Here, too, the high calcium content causes beads to form on the glaze surface.

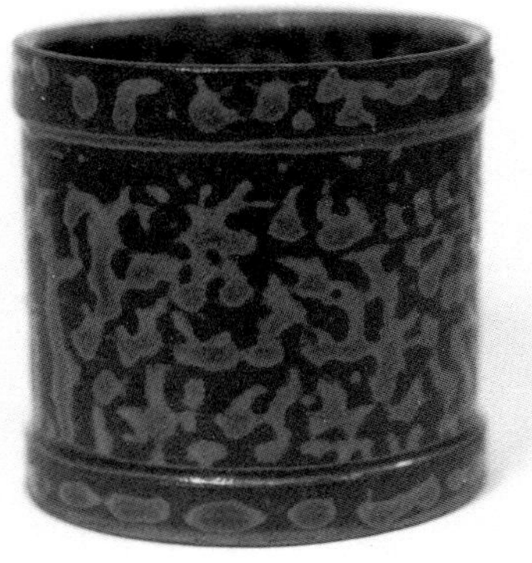

104. Irabo glaze

When ash was combined with a high-iron clay or stone in the right proportions, a brown-black glaze resulted. The first Japanese example occurs in fourteenth-century Seto wares. For the famous Satsuma iron-glazed wares (see pl. 22), the potters mixed ash with a ferrous sand or used a single, iron-rich "glazing clay." Some ferrous clays or stones will even fire to a glaze by themselves. Well-known Japanese examples are Mashiko stone and Kimaichi stone; a similar material in the United States, although now depleted, is Albany slip.

Feldspar and Ash Glazes

It was difficult to make a stable transparent glaze with the clays and ash available to the Japanese potter, and from the sixteenth century, when consumers were beginning to demand wares with underglaze painting, the development of such a glaze became imperative. A solution was found in feldspar, a by-product of the decomposition of igneous rocks, containing silica, alumina, sodium, and potassium. Stable at stoneware temperatures and generally low in metallic oxides such as iron, feldspar became the key ingredient in all transparent glazes.

Today, potters classify feldspars by their alkaline content. Varieties range from a potash feldspar that melts into an opaque, viscous glaze at about 1250 degrees centigrade to a soda feldspar that melts into a more transparent glaze at 1100 degrees centigrade. The nomenclature is, of course, arbitrary: all Japanese feldspars have both potash and sodium, although the sodium in potash feldspar is less than 1 percent. In place of feldspar, Japanese potters occasionally use *tōseki,* the porcelain stone mentioned in Chapter 2.

A stoneware glaze can be made of pure feldspar. The glaze of Shino ware, the famous sixteenth-century tea-ceremony ware from the Mino region, has until recently been considered a pure feldspar glaze (see pl. 7). Recent analysis of Shino glazes, however, suggest that they also contained a fair quantity of ash. The snowy white texture previously associated with high percentages of feldspar instead seems to have come from from a prolonged firing at a relatively low temperature of about 1200 degrees centigrade. In any case, recreations of this glaze have been most successful where a high percentage of soda feldspar is employed, one containing between 3 to 6 percent sodium. A coarsely ground feldspar is preferable: apparently a range of grain sizes encourages the characteristic Shino texture. Some dedicated potters in Japan even mine decomposing granite, remove its coarse feldspar grains by flotation, and use these in a fairly rough grind. Small portions of clay, up to 5 or 6 percent in all, and/or soda, which compensates for a low-soda feldspar, are sometimes added. The pink-red blushes that appear on early Shino ware also may be due to high alumina content. Another high feldspar glaze from the same period is the viscous white glaze seen on certain Karatsu tea bowls.

In the early seventeenth century, a deliberate combination of hardwood ash and feldspar, coupled with a more efficient firing, resulted in a true transparent glaze. Due to the more complete melt, light is reflected through the glaze from the clay underneath. This glaze shows underglaze decoration to its best effect. Every region had a preferred available ash for transparent glazes. For example, Seto used camellia, Satsuma used a variety of oak *(nara),* Kyoto and Arita used a hardwood called *isu* (see p. 130–31), Takatori used another variety of oak *(kashi),* Kutani used bush clover *(hagi),* and Shigaraki used walnut husks. But the ash now broadly used for transparent glazes is a mixed material called *dobai*. This literally means "earth ash," and the name apparently comes from the earthen kitchen stoves and hearths where so much of this ash was once produced. This common ash, as it is sometimes called in English, has a low (no greater than 4 percent) iron content, which gives it a slight yellow tint in oxidation and a pale blue-green color in reduction. Early transparent glazes at Karatsu and Mino were composed of about 60 to 70 percent feldspar and 30 to 40 percent hardwood ash.

Many kilns, including Kyoto, Satsuma, and Mino, adopted a simple transparent glaze composed of ash and feldspar. But these simple glazes were rather soft, prone to run, and had a short range of maturity. Before long, potters found that additions of the more refractory straw ash corrected these faults. An ideal balance is approximated by combining 60 percent feldspar, 30 percent wood ash, and 10 percent straw ash. The generally acknowledged lower limit for a feldspar glaze is 50 percent feldspar.

Straw Ash Glazes

Straw ash, particularly the rice straw ash used by Japanese potters, is high in silica, containing around 50 percent silica as opposed to 30 percent in hardwood ashes. Its most interesting role is as a main ingredient in opaque white glazes (see pl. 5; fig. 133). A good standard is 30 percent feldspar, 30 percent wood ash, and 40 percent straw ash. The excess silica does not dissolve, and these particles refract and diffract light so that the glaze obscures the body. An opaque glaze is the result. Depending on a number of factors, including glaze thickness, temperature, and the presence of phosphorus, this glaze may separate into white and transparent phases, which are opalescent, the prized effect seen on Jun wares from Song- and Yuan-dynasty China.

Straw ash, like other high-silica ashes, does not combust readily and is usually burned only to a carboniferous black. The ash from rice husks makes an even better white than straw ash does. Bamboo ash has been extensively used for this purpose as well.

The famous white and mottled white glazes of the western Japanese kilns Karatsu and Hagi are made with straw ash, although Hagi potters occasionally substitute porcelain stone for the feldspar to produce a distinctive white. Straw-ash glaze is particularly reactive over a high-iron body, an ochre slip, or an iron glaze. This exaggerates the mottling, an effect called *namako.*

COLORED GLAZES

As mentioned previously, early colored glazes, especially iron glazes, were a natural result of melting local clays and stones with ash. Another way of achieving colored glazes is to add increments of oxide colorants to the colorless glazes mentioned above. Three base mixes yield a wide range of effects:

Ash Base		**Feldspar Base**		**Straw Ash Base**	
feldspar	20 %	feldspar	60 %	feldspar	30 %
wood ash	70 %	wood ash	30 %	wood ash	30 %
straw ash	10 %	straw ash	10 %	straw ash	40 %

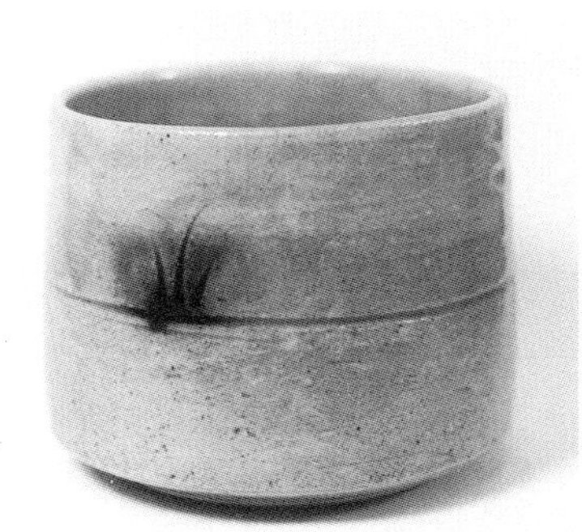

105. Yellow Seto

Many traditional iron glazes occur when between 1 and 10 percent iron is added to the above bases or variations thereof. Yellow Seto *(kiseto)*, a yellow semi-transparent stoneware glaze, was a favorite among tea masters from the sixteenth century (fig. 105). This is approximated by adding about 2 percent iron oxide to a modified ash base of 40 percent feldspar, 50 percent wood ash, and 10 percent straw ash. Some potters strive for a drier texture in this glaze, and that is accomplished by adding alumina in the form of 20 percent red clay or ochre, which will also supply the iron. The application should not be too thick. Regardless of the formula, a good oxidation firing is important, and the clay body should be low in iron.

Ame, an amber glaze, is achieved by adding around 4 to 8 percent iron to the feldspar base glaze. As the percentage of iron increases, the glaze becomes more fluid. This is balanced with increments of straw ash.

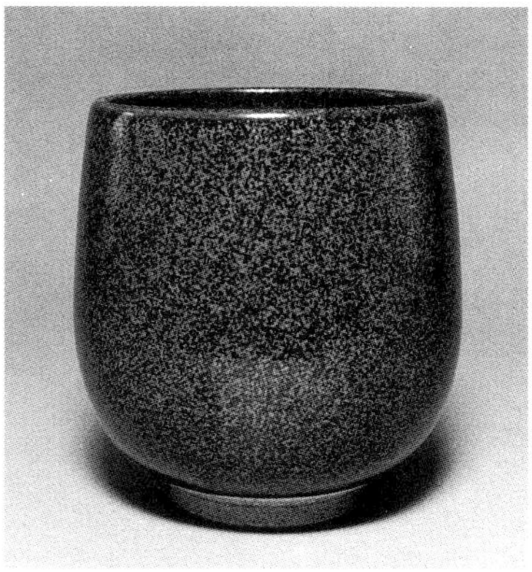

106. Soba

Soba glaze, named after the color of buckwheat noodles, is a mottled yellow-green color (fig. 106). It is basically an *ame* glaze devitrified by an excess of alumina. Increments of clay are added to the *ame.* Small increments of magnesia may contribute to the coloration.

Temmoku is a black glaze named after a mountain temple in southern China where a distinctive iron-glazed bowl was used in tea drinking. Many of these bowls were exported to Japan in the Kamakura (1185–1333) and Muromachi (1333–1573) periods, where they were prized by tea masters and also copied by potters, especially at Seto and Mino. The glaze is a black that breaks to a metallic brown where thinly applied. A modified feldspar base of 50 percent feldspar, 40 percent wood ash, and 10 percent straw ash, and about 10 percent iron oxide is appropriate. The quicker the cooling, the blacker the glaze will be.

Kaki is the bright orange-brown "persimmon" glaze made popular in the West by Hamada Shōji. This is similar to *temmoku*, but has less feldspar and more silica. A modified ash base of 25 percent feldspar, 40 percent wood ash, and 25 percent straw ash is combined with 7 percent iron oxide. About 2 percent titanium may also be added. Some potters recommend substituting a small amount of magnesium for the ash in the glaze. This is fired to at least 1250 degrees centigrade. *Kaki* works well when applied over a transparent glaze.

The other traditional colored glaze, Oribe, is pigmented with about 6 percent copper carbonate added to the feldspar base (see pls. 10, 11). The glaze must be fired in oxidation or it will turn brown or red. In the traditional Oribe ware, this glaze was applied to parts of the body, and the reserved areas were decorated with iron painting. Then the entire vessel was coated with a transparent glaze. The stiffer, more opaque green glaze used at such kilns as Agano and Mashiko is made from about 5 percent copper oxide added to an opaque rice straw ash glaze.

Through testing, it will be seen that each glaze exhibits its best effect at slightly different temperatures and atmospheres. Japanese potters realized this and used it to their advantage. For example, at Karatsu, the transparent ash glaze, which is "weak" or easy to melt, was packed in the cool spots to the rear of the kiln chamber, and the "strong" or refractory rice-straw ash glaze was packed in the hot spots at the front or top. In the late nineteenth century, Kyoto potters formulated their transparent glazes with 10 percent increments of limestone to match kiln conditions (cool spots, high lime) and underglaze decoration (crisp painting contours, low lime). In short, the standard glazes of any given workshop were closely related to the way in which their kilns fired, a point that will be discussed further in Chapter 7.

SYNTHETIC ASH GLAZES

Not everyone has access to copious amounts of ash. There are several ways of achieving the effects of ash glazes without using ash:

1) Mixing up batches of synthetic wood ash and synthetic straw ash and formulating on that basis. Ōnishi gives the following synthetic ash formulas:[1]

Synthetic Wood Ash	
potash feldspar	12.0
limestone	63.0
magnesite	5.0
bone ash	7.0
kaolin	10.0
silica	3.0

Synthetic Rice Straw Ash	
potash feldspar	22.2
limestone	5.8
silica	72.0

2) Instead of making synthetic ashes and then mixing them with feldspar or other ingredients, the entire glaze can be synthesized. Tagai recommends the following:[2]

Synthetic Wood Ash Glaze	
potash feldspar	54.3
talc	5.8
limestone	18.4
kaolin	2.9
silica	16.3
bone ash	1.4
iron	0.6
manganese	0.2

Synthetic Rice Straw Ash Glaze	
potash feldspar	48.2
talc	4.4
limestone	15.6
kaolin	0.2
silica	29.0
bone ash	1.0
iron	0.6
manganese	1.0

3) Or, one may make Western-style glazes and modify them. Three suggested base glazes are:

Transparent		**Opaque**		**Matte**	
feldspar	50 %	feldspar	50 %	feldspar	50 %
silica	25 %	silica	35 %	limestone	15 %
limestone	15 %	limestone	15 %	kaolin	35 %
kaolin	10 %				

Then, as Tagai has done in his formula, one may variously introduce increments of talc, bone ash, iron, and manganese. Talc supplies magnesia, which plays a prominent role in ash glazes as a flux, while bone ash supplies phosphorus, which will cause the opalescence characteristic of many ash glazes. Iron and manganese are present in small amounts in most ashes and hence they are included. Materials not mentioned by either Ōnishi or Tagai but important in most ashes are the soluble alkali, which can be supplied by soda ash (soluble sodium) or pearl ash (soluble potassium). Since they are soluble, they will promote a greater reaction upon absorption into the clay body.

4) If one has access to only a moderate amount of ash, it may be substituted for some of the standard ingredients in the glaze. For example, straw ash may replace silica in a glaze, or wood ash may replace limestone, since commercial whiting has only about 10 percent more calcium oxide than the average hardwood ash. The so-called ash glazes sold by Japanese supply houses contain an increment of ash added to a standard mineral glaze. The commercial ash glaze that I purchased in Kyoto was made of 60 percent feldspar, 20 percent wood ash, 10 percent limestone, and 10 percent kaolin. This could be further modified as in step 3, above.

GLAZES FOR PORCELAIN

Common wood ash generally has too much iron to be of use in producing pure whiteware. The ash made from the bark of the *isu* tree (a dark hardwood, *distylum racemoscum*) on the other hand, contributes to a colorless transparent glaze of less than 1 percent iron and fits porcelain bodies exceptionally well: there is less shrinkage of the glaze, and therefore less crazing, because of a higher silica content. The *isu* tree was at one time found in great quantities on the Ōsumi Peninsula in Kagoshima Prefecture, and the tree was burned exclusively for glaze. Seventy percent *tōseki* and 30 percent *isu* ash made a fine porcelain glaze. Compared to the present limestone glaze used for porcelain, it also had a slightly rough texture, which was advantageous for overglaze enameling, since enamels will not stick well to the smooth surface of a milled stone porcelain glaze. A contemporary equivalent for *isu* ash glaze is:

potash feldspar	34 %
silica	37 %
limestone	15 %
kaolin	13 %
manganese	1 %

LOW-FIRED GLAZES

The low-fired alkaline fluxes used in the West were unknown to traditional Japanese potters. Lead silicate glazes, composed of approximately 80 percent lead carbonate or oxide and 20 percent silica, constitute the principal low-fired glazes and were first used in Japan in the seventh century, when they were transmitted from the Asian continent. The first real domestic innovations in lead glazing, however, were carried out in the sixteenth century by two Kyoto workshops, Oshikōji and Raku. The Oshikōji workshop developed a series of pigmented lead glazes made from a transparent base glaze with small additions of oxides such as cobalt, copper, iron, antimony, and manganese, which it used for painting on bisqueware. These designs were then covered by a wash of the transparent base glaze and fired to a temperature of about 750 degrees centigrade—red heat. This produced a soft, diffuse effect akin to painting on absorbent paper.

The name Raku, originally taken from the Jūraku palace of sixteenth-century warlord and art patron Toyotomi Hideyoshi, is known to Western potters due to the American appropriation of the technique in the 1960s. The traditional Raku-ware style has two main variations, red and black (see pl. 12). The red is simply a transparent lead silicate glaze over an ochre-slipped body, but the black glaze is distinctive for its use of frit, a prefired glass. The Black Raku glaze is made of lead frit and Kamogawa stone, a ferromanganic rock from the Kamo River in Kyoto. The traditional proportion was 60 percent Kamogawa stone to 20 percent frit. Today the frits and glazes are more complicated. For potters interested in this type of glaze, it is best to seek out rocks like Kamogawa stone and melt them with various fluxes, preferably non-toxic alkaline ones.

The Marusan and Yumachi kilns in Fujina, Shimane Prefecture, on the Japan Sea also employ lead glazes. Their work, inspired in part by English slipware, makes use of a high-firing lead glaze composed of about 57 percent feldspar, 28 percent lime, and 15 percent lead.

I do not advocate the use of lead or lead glazes. Lead is toxic in its raw form, and lead glazes may also be toxic after firing, for they are soluble in organic acids such as orange juice. Lead glazes, therefore, pose a special hazard when used on the interiors of food or beverage vessels.

MIXING AND APPLICATION

Before mixing and applying glazes, it is important to decide whether to work with bisqueware or raw ware. Until the second or third decade of the seventeenth century, Japanese potters seldom bisque fired their wares. Even today many stoneware potters maintain that bisque firing impairs the thorough interaction of body and glaze that is central to the beauty of glazed stoneware. But with the development of thinly thrown forms and extensive underglaze painting (and, interestingly, with the introduction of feldspathic glazes, which fit better on bisque-fired bodies than on raw bodies) bisque firing became more common.

Glazes should be mixed and milled in much the same way as slip (see p. 105–106). The glaze slurry can be hand or machine stirred if the ingredients have been commercially milled, or ball milled if they are coarse. Traditionally, the stone portion was ground by hand (fig. 107). The glazes used to be filtered through silk mesh, but at present, 100- or 120-grade wire-mesh screens are used. Japanese potters use *funori* seaweed syrup in their glazes for suspension and adhesion. *Funori* also hardens the glaze coating and protects it during handling. The adhesive benefit is especially important when using ash glazes, which have little dry strength. The preparation and percentage addition of *funori* is the same as for white slip (see p. 106).

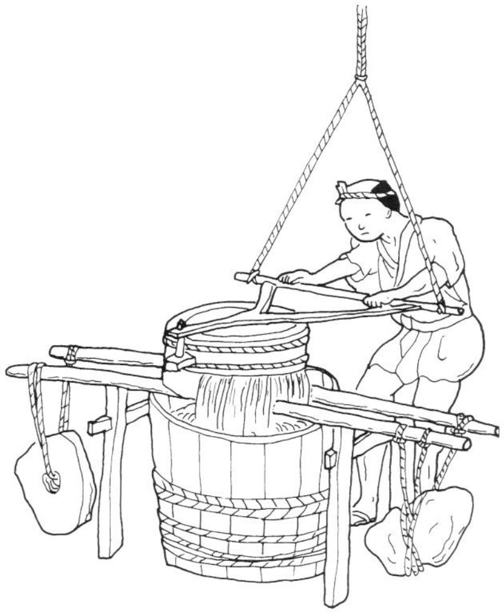

107. Glaze milling

Japanese potters rely on experience to determine how thickly to apply their glazes. The water content is judged by watching how quickly the glaze drips off a stirring stick. The thickness of the glaze coat is tested by dipping a piece of throwaway bisqueware into the freshly stirred glaze. After the glaze coat has dried, it is scratched through with a fingernail. Although there are many exceptions, the glaze coating should average between 1 and 3 millimeters. If the glaze coating is too thick, water is added. If it is too thin, surface water is decanted after the glaze has settled, or else the glaze is put into a large bisqueware vessel until a portion of the water is absorbed. It is harder to thicken a glaze than to dilute it, so glazes are mixed slightly thicker at the outset.

Thickening of the glaze coat around corners, foot rings and other places is avoided by brushing plain water on those areas just prior to application. The glaze will not thicken excessively in those spots.

Dipping, pouring, and blowing by atomizer or compressor are the main techniques for applying glazes. The first two have already been discussed in relation to white slip, and the third is a familiar Western technique.

Glazing cups (fig. 108). The Japanese potter has developed an efficient method for glazing small, deep objects in which the outside is first dip glazed, and the inside pour glazed. The bottom of the pot is held with one hand while the other scoops up some glaze with the ladle and pours it into the vessel. The amount here is important: one-third of the vessel capacity is about right (1). The glaze is poured out as the pot is rotated until the interior is completely coated (2). Any outside drips are cleaned up before going on to glaze the outside. If the bottom is to be left unglazed, the pot is held by the foot

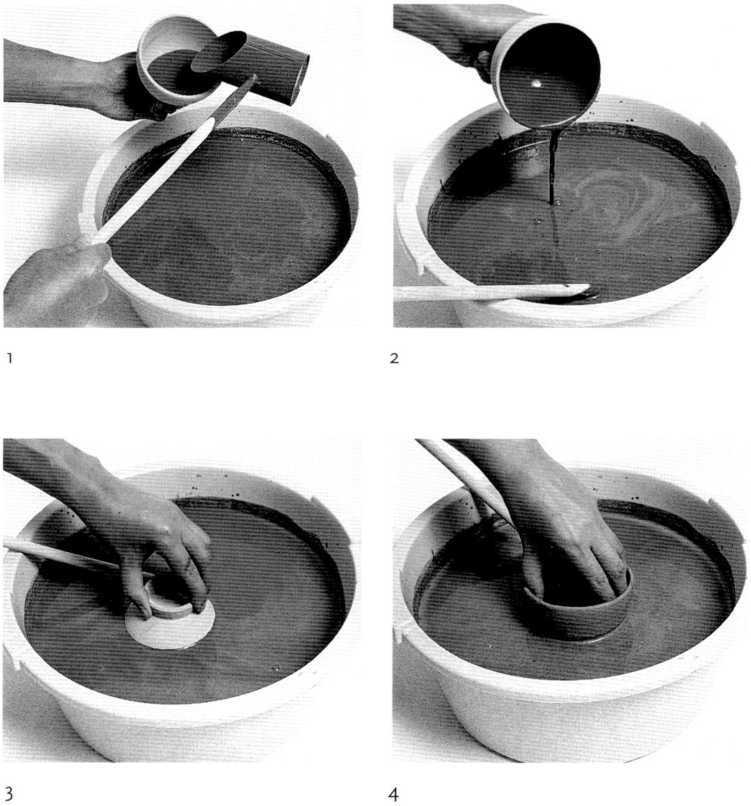

108. Dipping a teacup

and immersed into the glaze. The opening of the pot must be kept horizontal or glaze will bubble up into the inside (3). When the glaze coat has attained the desired thickness, the pot is pulled out of the glaze. When the entire outside is to be coated, the pot is held from the inside (the inside coat must have become sufficiently dry) and immersed bottom first into the glaze (4). The rim must be kept perfectly horizontal or glaze will pour in.

Wares that can be securely grasped by the foot are sometimes dip-glazed in a single procedure by immersing the inverted pot, jerking sharply upwards to suck glaze into the interior, and then swirling the glaze around the interior. It is necessary to work fast, and obtaining an even application is very difficult. This technique cannot be used with raw wares.

Dipping small flatware (fig. 109). For single-step dipping of plates up to 20 centimeters in diameter, the plate is grasped with the thumb on the rim and the index and middle fingers on the foot ring. Beginning with the rim opposite the fingers, the entire pot is immersed

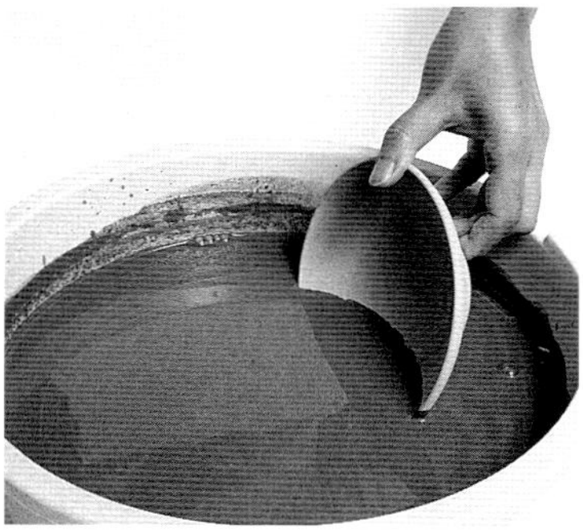

1

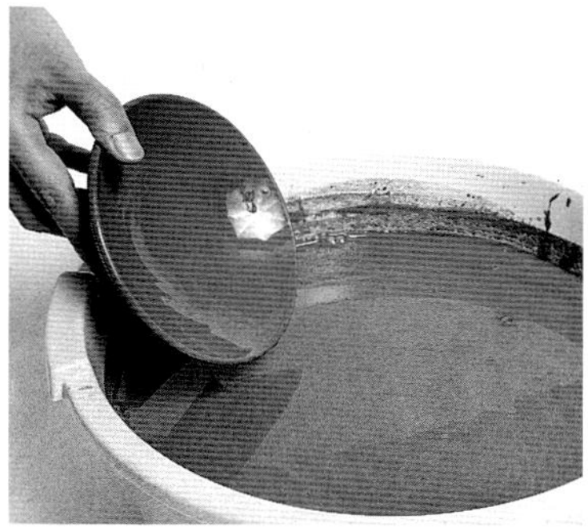

2

109. Dipping small flatware

in the glaze (1). Without changing direction, the pot is lifted out with the other hand (2). Bare spots are touched up with a brush.

Dipping large flatware (fig. 110). For plates larger than 20 centimeters in diameter, the pot is tilted over the glaze bucket. Glaze is ladled into the interior as the pot is rotated. The excess is allowed to flow back into the bucket (1). After the glaze dries, the vessel is turned over (2).

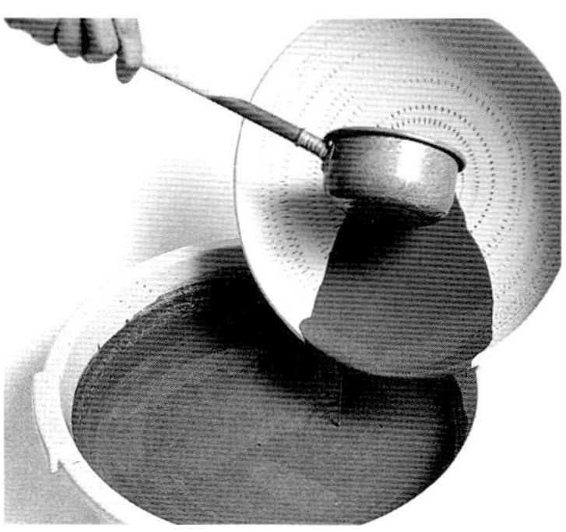

1

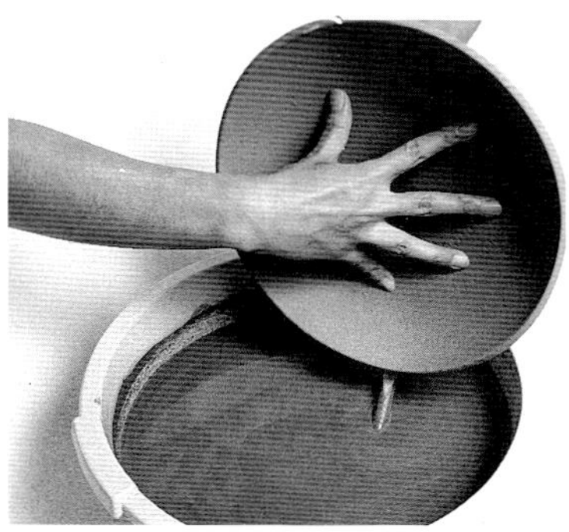

2

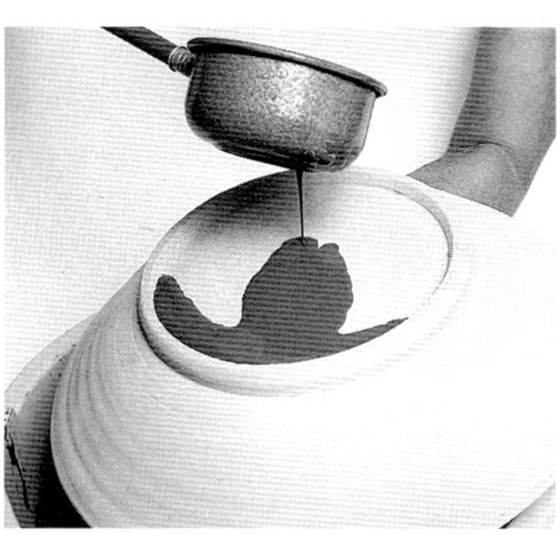

3

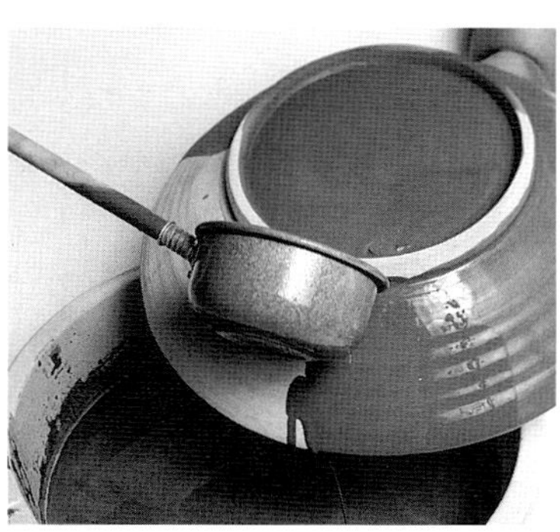

4

110. Dipping large flatware

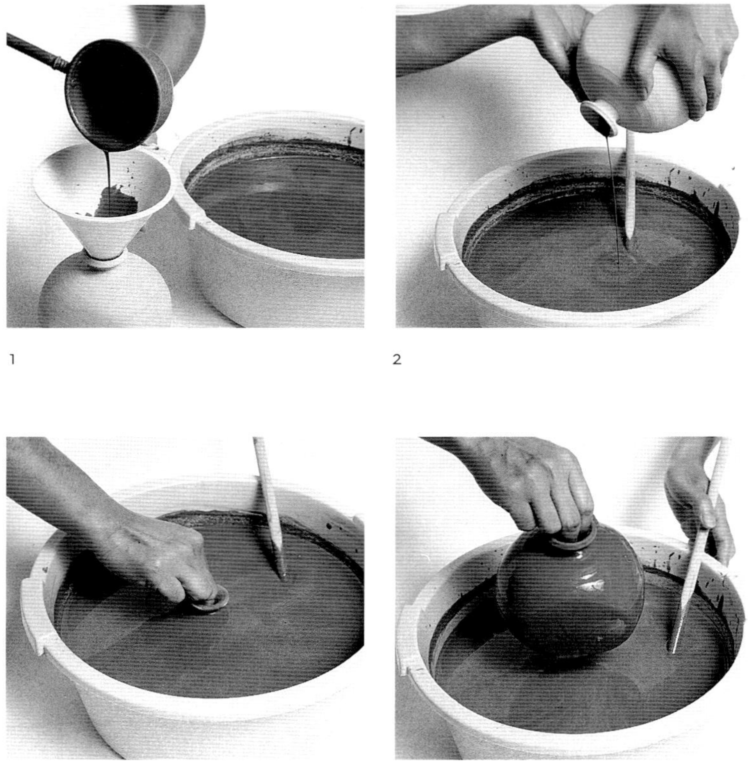

1 2 3 4

111. Dipping small hollow ware

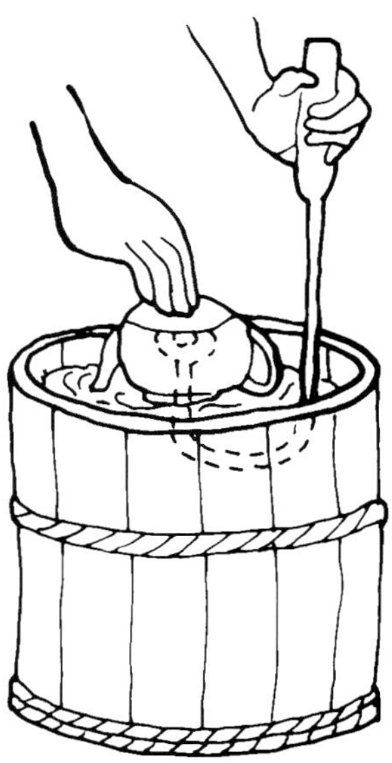

112. Dipping a teapot

Any glaze drips that have formed on the bottom are cleaned off. Glaze is ladled into the inside of the foot ring (3), and the pot is gently agitated until the glaze is distributed evenly. Any surplus glaze is then poured out. The pot is tilted over the glaze bucket, and glaze is ladled onto the area outside of the foot as the pot is rotated (4).

Dipping small hollow ware (fig. 111). For small bottles and the like, the interior is filled about one-quarter full of glaze using a funnel (1). The pot is tilted over the glaze bucket, and all the glaze is poured out as the pot is rotated (2). Exterior drips are cleaned off. If the fingers fit into the neck, they are used to hold the pot when it is inserted into the glaze (3). The pot is lifed out of the glaze (4). If the neck is too thin for such a procedure, the pot is held by the foot and inserted into the glaze upside down. Some workshops use an L-shaped tool for bottom-first dipping of bottles. The pot base rests on the tool, and a pointer provides pressure from the rim as the pot is dipped. If the pointer is removed quickly, the wet glaze will fill in the hole. There is also a specific dipping tool for use with wide-mouthed jars and teapots (fig. 112).

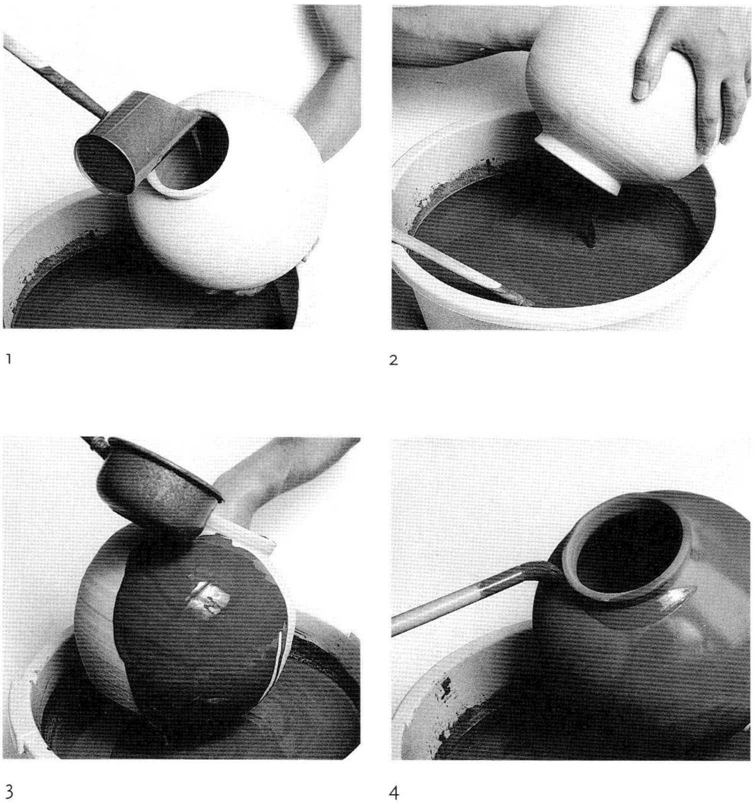

113. Glazing large hollow ware

Glazing large hollow ware (fig. 113). The inside of the pot is filled about one-quarter full of glaze (1). The pot is inverted over the glaze bucket and all the glaze is poured out as the pot is rotated (2). Large objects frequently displace too much liquid to dip into an average-sized glaze bucket, so instead, glaze is ladled onto the exterior as the pot is rotated (3). The neck will have some bare spots, and they are filled in with the brush (4).

Some cleanup is necessary after applying glazes. For example, glaze frequently splashes onto the foot ring or the base of the pot. This is cleaned when the pot is dry. It can be sponged off, but this tends to work the glaze into the pores of the clay. It is much better to use a scaling brush (see p. 40) or a hard toothbrush.

Decorating with Glaze

Double glazing is the most popular manner of glaze decoration and one well known outside of Japan. In this technique, the pot is usually given an overall coating using one of the techniques previously mentioned, and contrasting glazes are applied over that. There are several variations. One of the earliest methods, and by far the most common, is that of a rim dipped in a contrasting glaze. The Karatsu potters coated their bottles, jars, and vases with a dark iron glaze and then dipped the rims in a white straw ash glaze. This combination subsequently became popular at folkware kilns throughout the country. Only slightly less popular are trailed or poured patterns, which can be seen in Shōdai ware (rice-straw ash glaze dripped or ladled over brown iron glaze), Koishiwara ware (copper green glaze and iron glaze ladled over white slipped surfaces), and Ushinoto ware (rice-straw ash gaze trailed over iron glaze). Round "window" patches such as those seen in Tsuboya ware from Okinawa are also popular among dipped patterns. In the *donko* glazing of Satsuma, a coating of thick glaze, applied directly by hand, is covered by a glaze of a contrasting color. The latter breaks over the former in a pleasing mottled effect. Glaze accents may also be painted or inlaid, or a dry glaze may be sprinkled over a regular coating. (Wood ash or straw ash is usually sprinkled.)

Double glazing (fig. 114). After an initial glaze coat, a contrasting glaze is applied using such techniques as ladling (1) or dipping (2). It is important that the base glaze be still moist, or else the upper coat will flake off. *Funori* seaweed syrup is usually used to retard drying and promote adhesion.

114. Double glazing

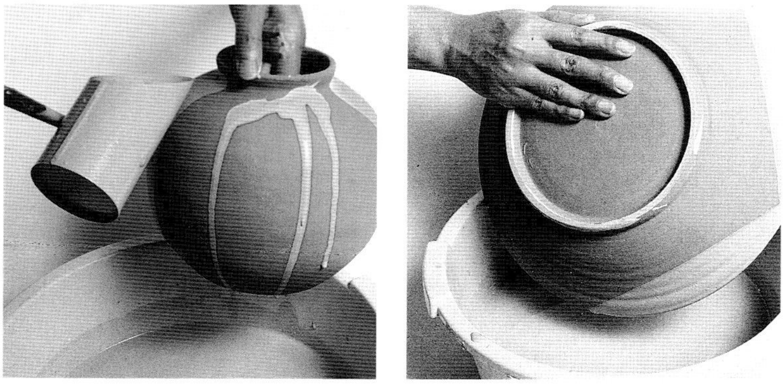

1 2

OVERGLAZE ENAMELS

Only a limited variety of colors is attainable in traditional high-temperature ceramics. From the mid-seventeenth century, the strength of high firing was combined with the colorfulness of low-fired glazes by painting the latter over glazed stoneware or porcelain and fusing them to the surface in an additional low-temperature firing (see pls. 15, 17, 18). Unfortunately, this enameling technique is infrequently practiced by Western studio potters, perhaps because of its association with the gaudy export wares of the nineteenth century or with china-painting hobbyists. Furthermore, it requires specialized skills and materials.

Nevertheless, enamels have great decorative potential. Not only are they appropriate for pictorial design, but they may be used for motifs or for abstract washes of color. When one begins to consider enamels as part of the entire process of making a pot, and not just as a decorative afterthought, that potential can be realized. The "ceramic" nature of Japanese enamels is revealed in their formulas—most of them fire thick and glassy. In contrast, traditional Western enamels are saturated with colorant and lie flat on the surface like a paint.

Frit *(tama, shiratama,* or *biidoro)* is the main ingredient in overglaze enamels. As was mentioned above, frit is a prefired glass. The ingredients (in the Japanese case, silica, lead, and sometimes a little borax) are placed in a crucible, heated into a glassy solution in the kiln, cooled by sudden quenching, and then ground into powder. Frit has a number of advantages in an enamel. It lessens toxicity, it facilitates even maturation by allowing gasses to escape in the prefiring, and it prevents soluble materials from being absorbed into the body. It is more durable after firing, and will stand up to wear. Ideally, a single frit and colorant should be enough to formulate an enamel, but each colorant affects the melt differently. In order to get different colored enamels to melt at the same temperature, it is necessary to make adjustments to each enamel: those that mature early are supplemented with silica, which acts as a refractory "brake," and those maturing late are given added flux in the form of lead carbonate.

Before discussing the various colors, it is important to consider whether it is at all worthwhile—or safe—to work in the traditional lead-based enamels. They do yield incomparably rich colors, but this author thinks that the danger involved in the use of lead outweighs any advantages. It is probably better to grasp the underlying principle of each color and experiment with nonleaded frits. Clearly, some testing will be necessary.

Enamel Colors

Red: The color comes from very pure iron oxide, traditionally obtained by baking iron sulphate. Commercial iron oxide, a by-product of separating aluminum from bauxite, is generally too impure to yield a good red color. In contrast to other Japanese enamels, which are almost entirely lead silicate glass, the red enamel must have a high percentage of colorant, or else it will dissolve. A suggested base is 50 percent frit, 25 percent flux, and 25 percent iron. The latter must be extremely finely ground. The red is applied thinly and takes its gloss from the glaze underneath.

Green: The color comes from copper oxide, which in the past was obtained by oxidizing scrap copper in salt water, and then thoroughly washing it (see p. 119). Modern copper oxides produced by electrolysis turn a blackish green. A suggested base is 50 to 60 percent frit, 20 to 35 percent flux, and 6 to 10 percent copper, preferably carbonate rather than oxide.

Yellow: A very soft yellow traditionally has been derived from antimony oxide. But antimony is both toxic and soluble, so in this case it is better to work with nontoxic commercial stains. Iron may also serve as a colorant in yellow enamel, but this imparts an orange hue.

Blue: The traditional colorant was a blue frit known in the West as smalt. The color was probably fritted because cobalt tends to bubble when fired raw. Even when diluted in a glassy enamel, commercial cobalt may be too intense, so the Japanese used a natural cobalt (see p. 114) which worked well with other colors. A suggested base is 60 percent flux, 20 percent frit, 15 percent silica, 4 percent cobalt (this may have to be calcined first), and 1 percent iron.

Purple: This was the most difficult color for traditional enamelers. Most of the Japanese material came from a high-manganese asbolite (natural cobalt or *gosu*) mined on the Noto Peninsula in

Ishikawa Prefecture. A suggested base is 46 percent frit, 46 percent flux, and 5 percent silica. Coloration may be variously provided by up to 5 or 6 percent additions of manganese, natural cobalt, and perhaps a dash of potassium permanganate. In any case, the color is difficult to achieve.

Black: This is another difficult color. The traditional black is actually composed of two applications: first, a black pigment consisting of about 50 percent manganese, 35 percent natural cobalt, and 15 percent silica; and second, a translucent enamel, usually yellow or green, painted over the black. This double application, seen most frequently in Kutani ware (see pl. 17), was necessary because it was impossible to achieve a good black color and a gloss in a single formula.

Preparation

It is helpful to have a power mortar, since these materials take hours of grinding, especially if obtained in the raw state. The ingredients are ground until the consistency is smooth and even. One traditional labor-saving device was a mortar with pestle support (fig. 115). All enamels are diluted with *funori* seaweed syrup. The one exception is red. Since *funori* tends to disperse the iron particles in the red enamel, red is diluted with water alone or with *nikawa*, a glue made from boiling animal hides or horns. To prepare the glazed pot surface for enameling, it is first covered with a thin glue solution *(nikawa* is also used here). This treatment helps the enamels adhere to the surface during application.

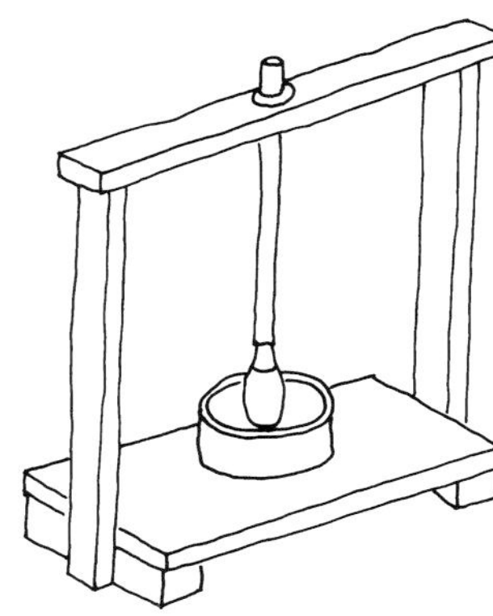

115. Mortar rack

When painting, a glass palette is used (one can be made by scouring a sturdy window-pane with emery cloth), and the enamels are periodically reground with a small, cube-shaped porcelain pestle. A different pestle, palette, and brush are used for for each color. These are stored without washing and rewetted at each successive use.

Firing

Traditionally, enamels were fired in a small, wood-fueled updraft kiln nearly identical to a bisque kiln, except that the enamel kiln had an interior muffle to provide more protection (see fig. 117). Since these kilns varied in design, different effects were achieved in each locale. For example, the famed Kakiemon reds of Kyushu (see pl. 17) were fired in a clear, oxidizing flame, which kept the reds bright. The Kutani enamelers had to fire their kilns in reduction to bring out their characteristic purple, so their reds turned a darker color. Now electric kilns are used everywhere. The kilns are fired for a period of 5 to 6 hours. Maturation is judged by draw tests. If the enamels are pitted at maturity, they may need further grinding, or the individual oxide colorants may need prior calcining. If a powdery salt precipitates at the margins of fired enamel, the process can be stopped by placing the pot in water and bringing the water to a boil. Red enamels are sometimes polished with very fine abrasive to deepen the color.

NOTES

1. Ōnishi Massataro, *Togei no Yuyaku* [Ceramic glazes] (Tokyo: Rikōgakusha, 1981), 3–23.

2. Tagai Hideo, *Nyūmon Yakimono no Tekuniku to Kanshō* [An introduction to ceramics technique and appreciation] (Tokyo: Kyōritsu Shuppan, 1979), 93.

SIX

Kilns and Firing

At first glance, there is little to distinguish traditional Japanese kilns from the wood-burning kilns developed earlier in China or Korea. Indeed, both low-firing earthenware kilns and high-firing kilns for stoneware and porcelain can be linked to continental models. Unique to Japanese kilns, however, is the way they are utilized. Traditional potters rarely attempted to achieve hair-trigger control as in the firing of copper-red glazes in China nor were they obsessed with sealing off the wares from combustion products. Instead, they devised methods to channel uneven heating, ash, carbon, and water vapor in ways that would decorate rather than mar the wares. This should not be seen, however, as a pragmatic bow to inadequate technology. Rather, potters found in the marks of the fire something essentially ceramic and they strove to further master and define that essence. Even today, with the availability of cheaper and cleaner fossil-fuel and electric kilns, a reverence for wood firing continues unabated.

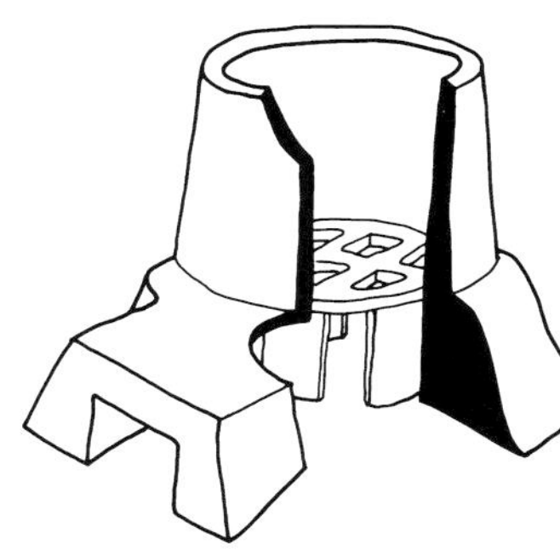

116. Low-firing kiln

TYPES OF KILNS

In prehistoric times, potters stacked their wares on open ground or in simple pits and covered them with combustible material, which was then ignited. A small kiln used for firing lead-glazed earthenwares was introduced from the continent in the late seventh or early eighth centuries. This was the prototype of later low-firing kilns. Whether they are for bisque firing, glazed-earthenware firing, or enamel firing, all low-firing kilns are basically the same. They are usually cylindrical updraft devices with a projecting firebox (fig. 116). Wares are stacked from the top. When this type of kiln is used for bisque firing, the entire inside is filled with greenware and in place of a lid, a layer of shards is placed over the top. Glazed earthenware and enamel kilns are lidded, and have an interior muffle (fig. 117) made from coarse fireclay, which protects the ware from

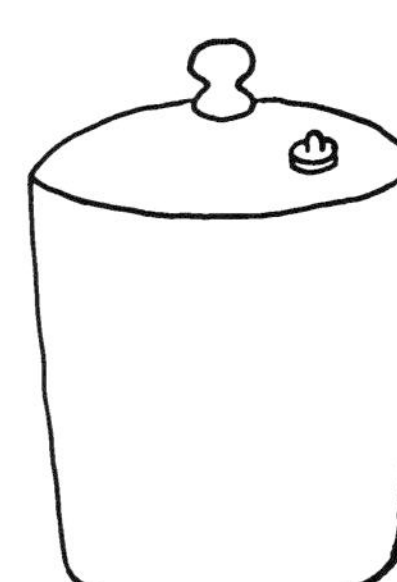

117. Muffle

strong drafts and ash. Basic firing time for any of the above is 5 to 6 hours to red heat (750–800 degrees centigrade).

The sloping tunnel kilns (*anagama*) used to fire Sue wares from the late fourth or fifth centuries were the first high-firing kilns. These consisted of either excavated tunnels or roofed trenches situated on hillsides. Wares were stacked in the open without shelves or saggars and fired in a crossdraft. The kilns could achieve temperatures as high as 1100 degrees centigrade, sufficient to partially vitrify the wares. An excess of fuel introduced at the end of the firing sealed the pots with carbon. A modification to the tunnel kiln, this time devised by the Japanese, appeared in the tenth century (fig. 118). It featured a dividing pillar, which facilitated better heat circulation and expanded capacity. A larger fire mouth allowed temperatures to rise as high as 1240 degrees centigrade, sufficient to melt an applied ash glaze. Since it can produce very spectacular effects and is relatively inexpensive to build, the tunnel kiln has undergone a revival since the 1970s.

In the next phase of development, kilns were constructed rather than dug out. The so-called "great kiln " (fig. 119), which was first used at Mino and Seto at the beginning of the sixteenth century, was a single-vaulted, single-chambered structure built almost completely above ground and supported by a row of interior pillars. Access to the kiln chamber was no longer through the fire mouth but instead through a side door. The fire mouth was much wider and opened into a bag wall made of clay pillars with saggars stacked on top. The flame was directed first upward by the pillars and then horizontally through the rest of the chamber, which was also filled with loaded saggars. A more even heat distribution greatly facilitated the production of glazed wares.

Almost a century after the inception of the great kiln, potters in Karatsu began using a single-vaulted, multichambered climbing kiln, the so-called split-bamboo style (fig. 120). The ware was stacked without shelves or saggars and fired in a series of chambers separated by vertical walls, perforated by flues at the bottom. A variant of the split-bamboo kiln is the "rifle" or "snake" kiln, which can still be seen in Tamba (fig. 121). These kilns fire very quickly. In the rifle kiln, pillars separate the chambers instead of walls. The pots are commonly stacked on ring-shaped setters without shelves or saggars.

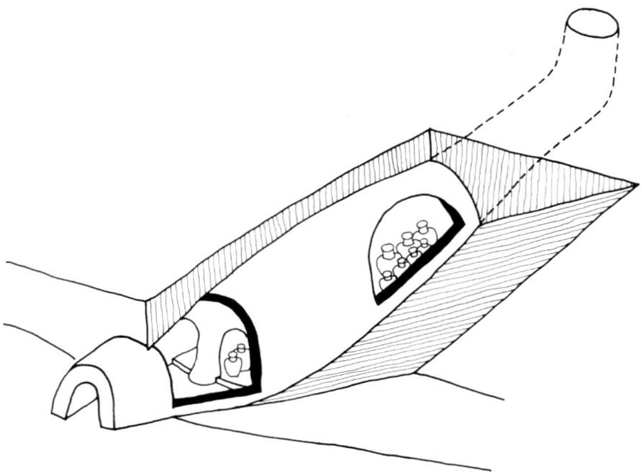

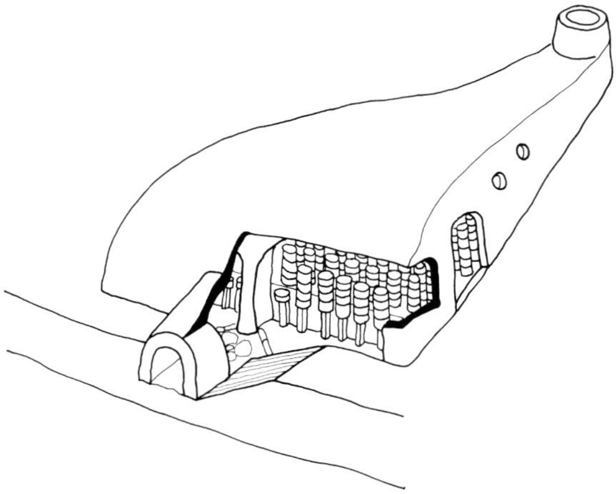

118. Tunnel kiln with dividing pillar *(left)*

119. Great kiln *(right)*

The mature Japanese climbing kiln is the multichamber, multivaulted climbing kiln (fig. 122). It was adopted from China or Korea at the beginning of the seventeenth century. The wares were stacked in a succession of stepped chambers, using shelves and saggars. As the flame hit a step, its path was diverted upward, then pulled back down though the flues at the back of the chamber. This combination of draft directions provided more even heat than earlier kilns allowed.

Most of the early kilns had wide, low chambers, which favored oxidation firing, but by the nineteenth century, many kilns developed taller, narrower chambers. This shape improved reduction firing, which was necessary for fine blue-and-white porcelain and, later, for special glazes like copper red. The multivaulted climbing kiln continued as the standard high-temperature kiln in Japan until just three decades ago. But the convenience of fossil fuels, rising firewood prices, and air pollution ordinances have encouraged potters to switch to fossil-fuel kilns instead. Very few multivaulted climbing kilns are now built in Japan. Most studio potters with an interest in wood firing build small *anagama*-style kilns, although most of them are built above ground from firebrick and have chimneys. These are typically long tubes or a single bulging chamber (fig. 123).

BUILDING A KILN

The site for a wood-burning climbing kiln is selected carefully. The preferred location is a sunny slope, facing south. Moist locations are avoided. The average wind direction is important: it should blow in

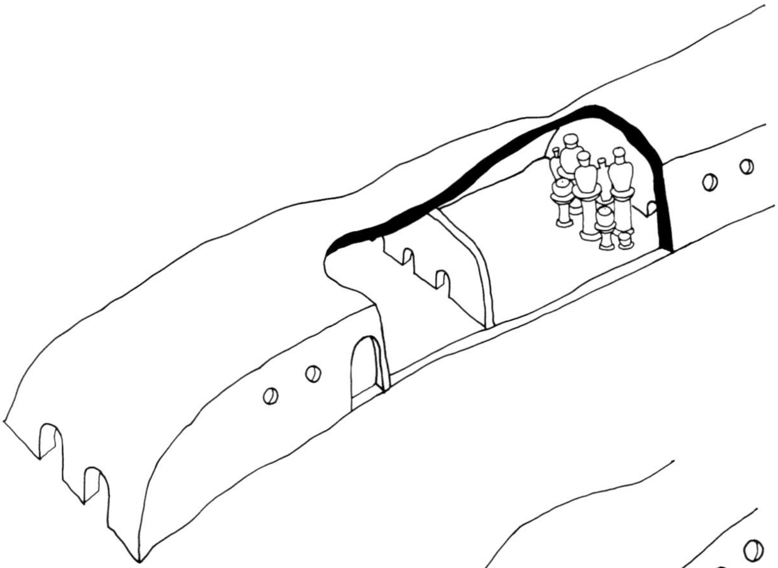

120. Split-bamboo kiln

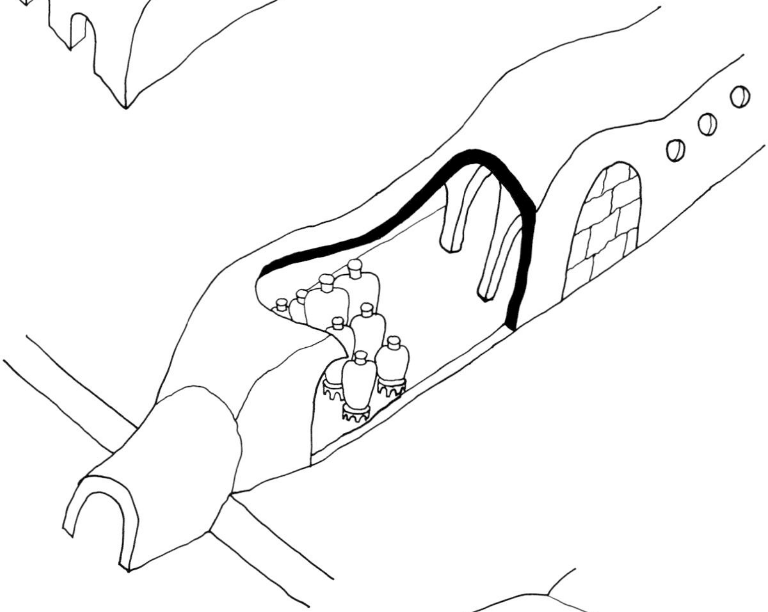

121. Rifle kiln

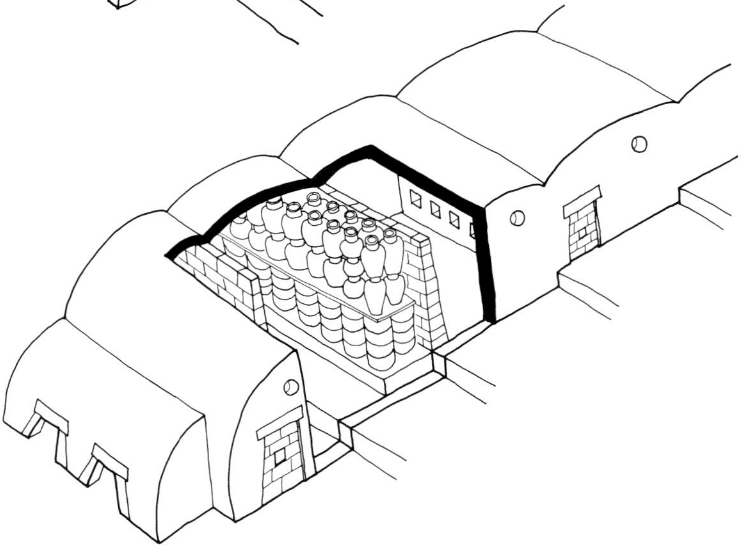

122. Multichamber, multivaulted climbing kiln

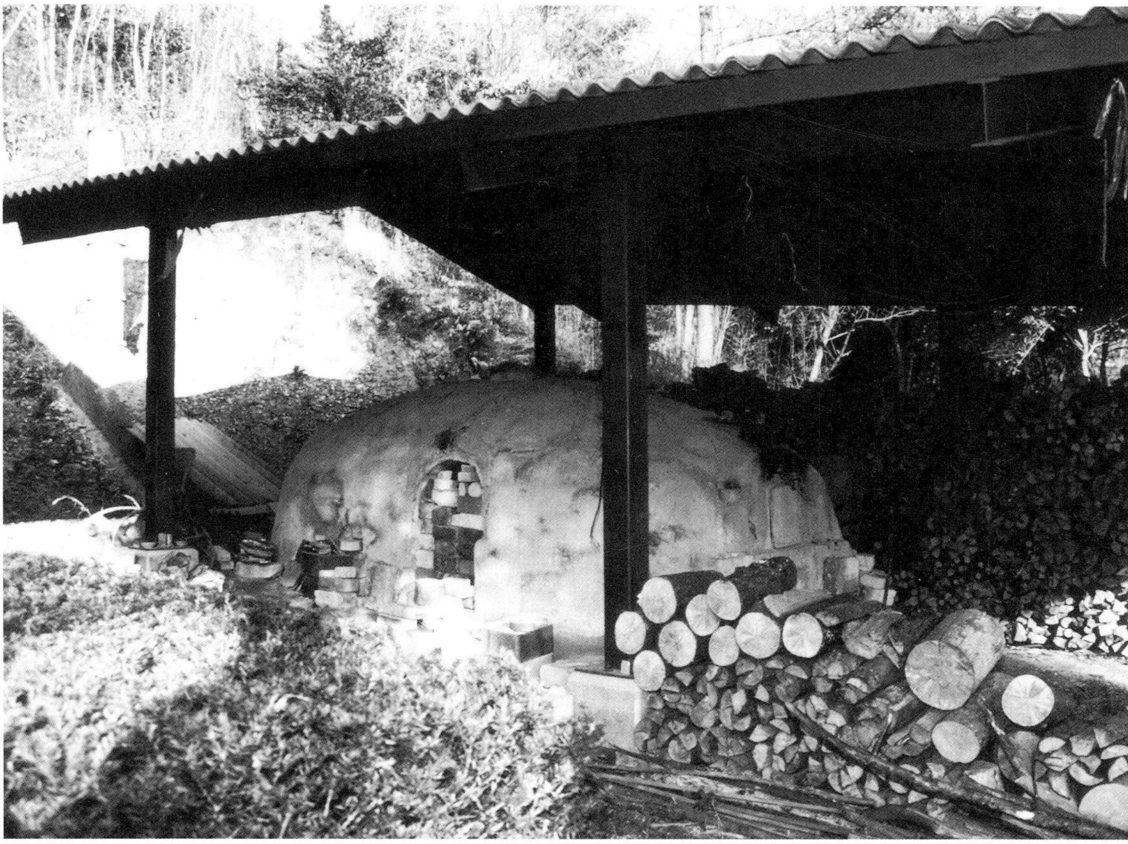

123. Contemporary "tunnel" kiln

the general direction of the draft. The slope is critical in determining the way the kiln will fire. Generally, the steeper the slope, the faster the firing. Japanese climbing kiln slopes range anywhere from 15 to 35 degrees.

The preferred time for kiln building in Japan is early autumn. The materials dry at an even rate at this time of year and there is no danger of freezing or excessive rain. Early spring is the best alternative. The plan of the kiln is staked out, the whole site roofed in, and the foundation is dug.

Refractories are then prepared. Until quite recently raw fireclay was typically mixed with a grog made of used refractory material. Repeated firings produce long needles of mullite, an alumino-silicate crystal that is extremely stable at high temperatures, in the microstructure of this grog. This clay/grog mixture either could be made into bricks or used as mortar. Refractory stones such as coarse sandstone and granite were occasionally used instead of blocks. Today most woodburning kilns are made from refractory firebrick.

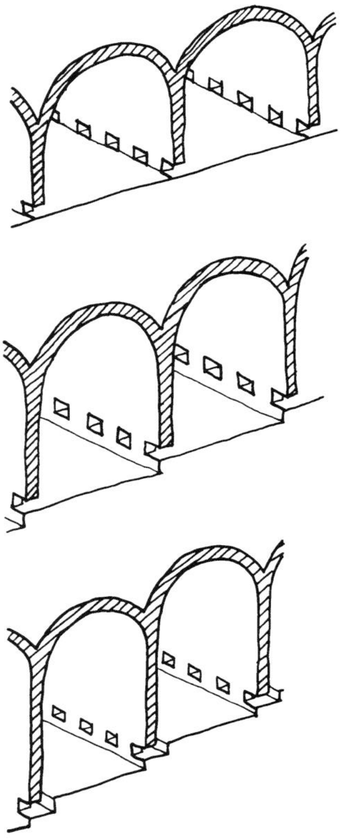

1

2

3

124. Flues

When the construction materials are ready, the walls between the chambers are built of high-refractory blocks about 30 centimeters thick. These partitions contain flues through which the draft passes (fig. 124). In multi-chambered climbing kilns, great attention is paid to the design of these internal flues. To compensate for lower heat along the sides of the kiln, the left and right side flues may be built wider, and the middle flues where the draft is typically hotter may be built narrower. Initially the flues are built oversize, since they are much easier to reduce in size than enlarge if modifications become necessary.

The relationship between the rise of the kiln and the flues is important. In the most primitive climbing kilns, the floor simply ran along the slope, and the flues followed that incline, resulting in so-called diagonal flues (1). As the kiln floor became leveled and stepped for convenience in stacking, kiln builders explored two options in placing the vertical step: either in the rear of the chamber, where the draft can travel uninterrupted into the next chamber (horizontal flue; 2), or in the front of the chamber, which makes the draft shoot upwards upon entering the chamber (vertical flue; 3). For example, the early Mino climbing kilns used the horizontal type, which fires quickly, while the later Kyoto kilns used the vertical type, which fires more slowly.

After the partitions are complete, the lower part of the outside wall is built, and then the arch frame, a basket construction of bamboo or saplings tied together with rope, is put into place. The upper wall, together with the requisite door arches, spy holes, and blowholes, is built, followed by the roof. A uniform thickness is important for both walls and roof. Before the introduction of arch bricks, either a stiff clay/grog mixture was pounded into place, or special cone-shaped bricks, called *tatami* bricks, were fitted and mortared into place.

The firebox is usually built last. The ash pit and grate of the firebox are constructed, followed by its front, sides, and vaulted roof. The front is built with generously proportioned single or multiple fire mouths, since it is easier to constrict them than to enlarge them later. High-refractory brick is used.

Most traditional Japanese climbing kilns do not have protruding chimneys or dampers. There is simply a narrow passage behind the flues in the last chamber, and the exhaust exits from its open top. This passage is the same height as the chamber.

About two weeks after the kiln is built, the interior supports are removed, and the whole structure is allowed to dry for two months. Some potters first fire it empty, or with empty saggars. In the old climbing kilns, the interior of the chambers was lined with a fusible clay, which helped to keep refractory debris from falling into the ware. A professionally built climbing kiln was expected to last between 100 and 200 firings.

LOADING A KILN

This is a daytime job. The loading depends on both the kind of kiln and the kind of wares. For example, an unglazed stoneware firing is different from a glazed stoneware or porcelain firing. In the case of unglazed stoneware, wares are usually stacked without the aid of saggars. Shelves may not be used either, in which case the wares are stacked mouth-to-mouth, small pots within larger ones, and so on. With unglazed wares like Bizen, which depend entirely upon kiln effects, much attention is given to placement. Each pot must be correctly stacked in relationship to the draft and to neighboring pots. A general rule is open stacking in the front chambers of the kiln and denser stacking in the back. Potters who work in unglazed stoneware say they think about the placement of the pieces as they are making them on the wheel. It is too late to think about it during loading.

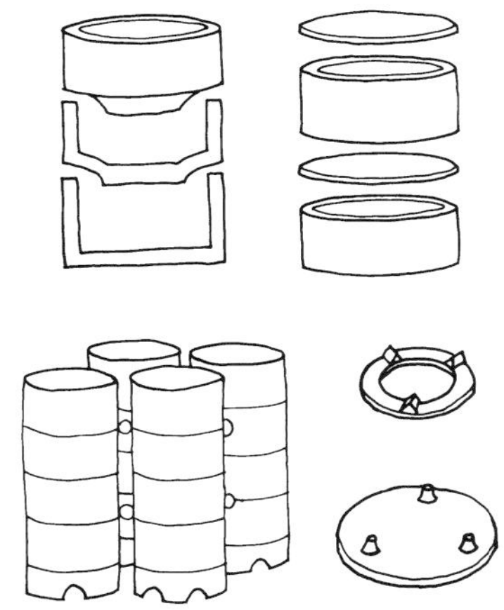

125. Saggars and setting tools

For glazed stoneware, the wares directly exposed to the flame are generally protected by saggars (fig. 125). These are stacked with the bottom of each saggar forming a lid for the one below it. Usually stacks of saggars are put across the front of the chamber, forming a bag wall that protects open shelves of wares in the back of the chamber. The distance between saggar stacks, about three finger-widths, is maintained by wedging balls of fireclay between the stacks. Shelves include the usual slab style with corner posts and in some cases, shelves supported by a single central pillar (fig. 126). The setting is hard work and calls for patience and care. Clay surfaces fuse

126. Open post stacking

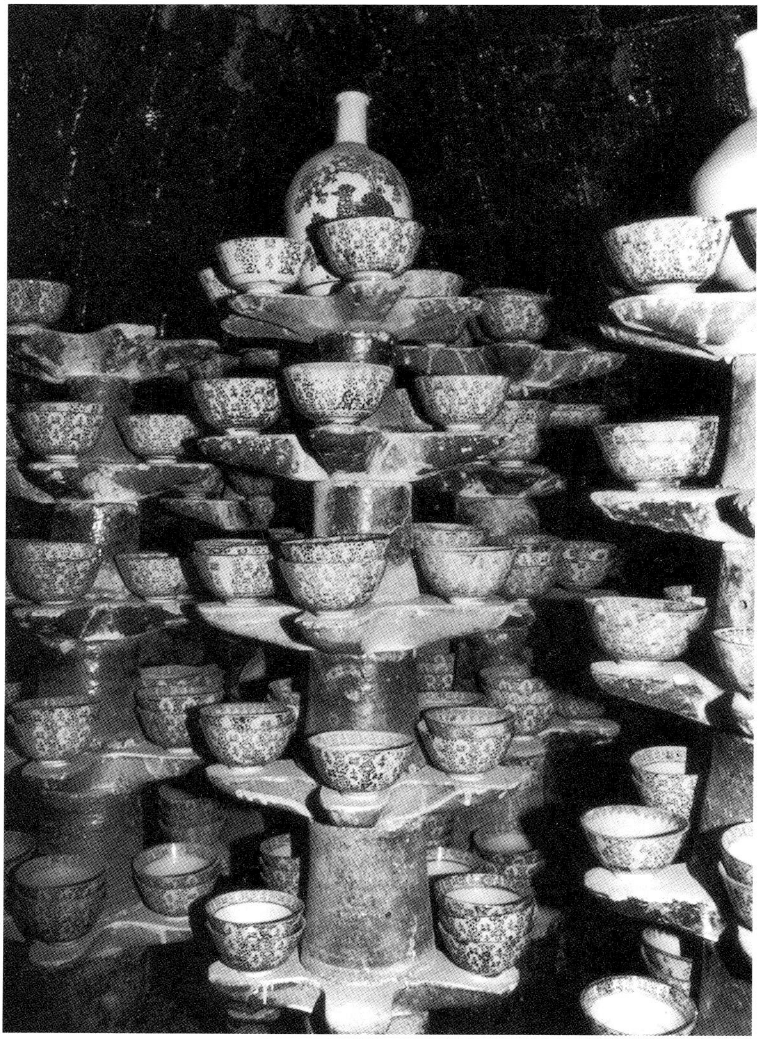

easily at high temperatures, and pieces have to be separated from shelves or saggar-bottoms with wads of fireclay. Some potters use sand or high-silica rice chaff to keep the pots from sticking to the shelves. Fireclay spurs and pads (see bottom right of fig. 125) are also used. Careful setting not only reduces loss, but also saves laborious cleaning after the fire.

The front of the climbing kiln usually reduces, and there is a great deal of flame and ash in circulation. This is the best place for natural ash-glazed ware. Wares with smooth glaze effects cannot be set in the open here. With saggars, however, glazed wares that need reduction, such as blue-and-white and copper-red, are usually fired in the lower chambers. The middle chambers of the kiln are more evenly heated, and some glazed wares can be set out in the open. White-slipped wares develop well in this part of the kiln. The rear chambers yield less interesting kiln effects, so wares with opaque colored glazes are often set there.

FIRING A KILN

Kiln fuel is usually softwood. For pure calorie value, the best woods are cedar, pine, and cypress, in that order. Length of flame and thoroughness of combustion, however, are more important than calories. Pine, particularly Japanese red pine, which produces a flame up to five meters long, is the overwhelming favorite, followed by cypress and cedar. It is said that pine cut between August and February produces the most calories. Diseased or bug-infested pine does not fire nearly as well as healthy pine.

The firing is traditionally conceived of in two stages, firebox stoking and chamber stoking. The former drives the moisture out of the kiln and brings the first one or two chambers up to red heat; the latter brings the ware to maturity. The kiln is loaded during the day, the doors are clammed up, and the firebox stoking proceeds throughout the night. During this stage, logs are stuffed into the main firemouth. The type of fuel for the preliminary heating is not as important as the individual chamber stoking fuel, and some potters use kerosene, oil, or wood debris. Naturally, the duration of this early stage depends upon the design and size of the kiln as well as the kind of wares being fired. In general, the longer the preliminary heating, the better. This is especially true for wares like Bizen, which are prone to break during fast temperature rises.

When the second or third chamber has achieved red heat, the chamber stoking phase begins (fig. 127). The main fire mouth is closed, and split wood, sticks about 60 centimeters long and 10 centimeters in diameter, is fed into ports on either side of the lowest chamber. The feeding must be very precise. If the wood is tossed in carelessly, the fire will choke. Veteran stokers say that wood must be inserted so that it will flip end-over-end and thus come to land in irregular patterns, keeping the flues open. The timing must also be

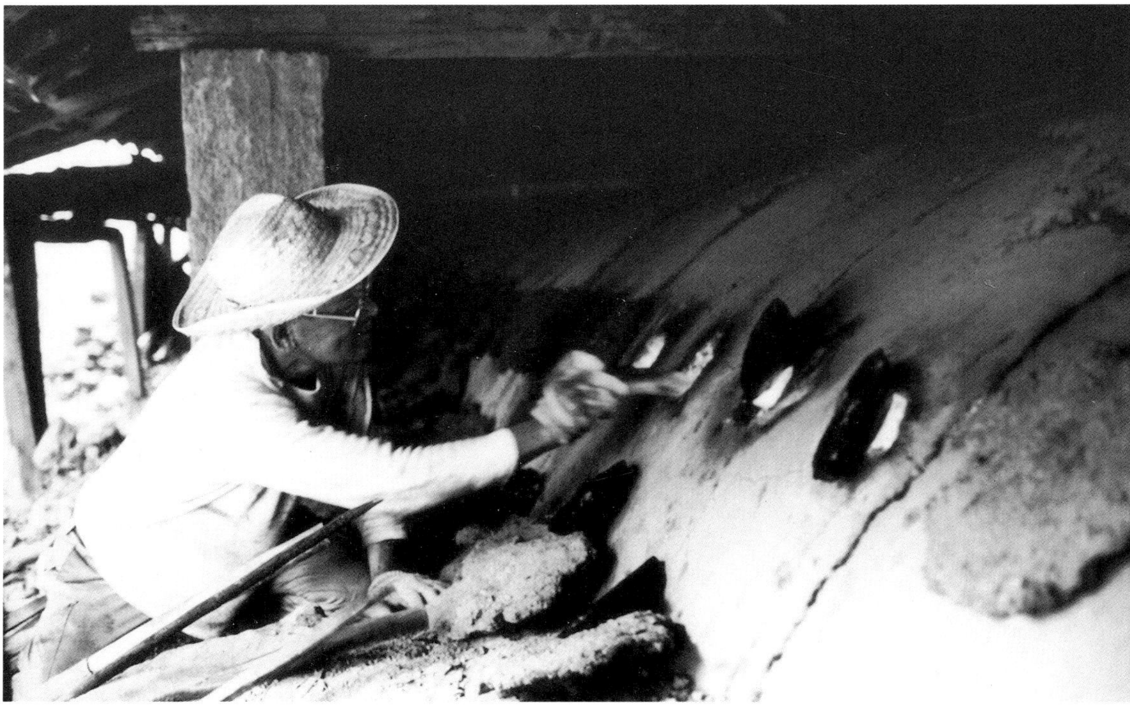

127. Firing a kiln

perfect. If too much time elapses between stokings, the temperature will decrease, but if too little time elapses, the draft will stall. Even for veteran potters, wood firing is a battle with the elements. Every imaginable factor can influence the fire: prevailing temperature, humidity, wind direction, the condition of the fuel, and even the stoking team. Months of work are at stake, and the most critical decisions must be made when fatigue is approaching a peak. As one potter related, "I feel as if my very soul is in there with the pots."

The goal during the chamber-stoking phase is to achieve an even temperature rise and even heating throughout each chamber. An imbalance of heat may be remedied by extra stoking, but some potters control this by opening blow holes built into the upper side of each chamber to pull the draft in that direction.

As one chamber matures in temperature, the stokers move to the next. This must be done quickly, with all unused openings closed promptly. Many potters still refuse to rely on pyrometric cones or pyrometers to judge the moment of maturity. They either read the kiln color, look at the gloss on the pots, or use draw tests. The kiln shows a succession of colors—red, orange, yellow, and white, and even purple—over the course of the stoking phase. The shape of the flame changes too. At peak temperature, around 1300 degrees centigrade, the flame currents appear long and threadlike. When the kiln is a little cooler, the flames are short and needlelike. Experts advise reading the color of the pots, not of the surrounding atmosphere, because the pots are usually one color stage behind the atmosphere. Maturity may also be judged by draw tests. The test pieces are small and usually perforated. They are set on a shelf or saggar near a spyhole. At peak temperature, they are hooked out with a metal rod and quenched in a bucket of water. Because of the quick cooling they will not be a reliable gauge of color, but they will show the glaze melt. Needless to say, common sense and long experience are necessary to fire the kiln successfully and to know when to stop firing.

After the firing, the kiln cools for about the same time it took to achieve peak temperature. Then the wares are unloaded, the shelves and saggars are stacked, and the kiln is cleaned up. Small burrs and scars on pieces are cleaned with old sandpaper, and the pots are washed. Wood-fired, unglazed pieces are frequently coated with saké or oil to deepen their coloration.

FIRING SCHEDULES

Traditional wares were fired without the use of precisely timed schedules or modern techniques of kiln atmosphere control. Potters coordinated clays, decoration, glazes, and loading techniques with the kiln effects that they wanted to produce. But with the ascendency of fossil-fuel kilns, potters have given much thought to ways of achieving special effects through control of the kiln atmosphere. Instead of setting up schedules where oxidation or reduction is continuous, they have found that intermittent schedules contribute toward distinctive effects. Manipulation of the cooling schedule and atmosphere is also important.

To achieve early reduction, the kiln is fired in oxidation to 900 degrees centigrade, then in reduction from 900 degrees to 1100 degrees centigrade, followed by oxidization to maturity. This is effective for the pinkish halo effect seen on Kyoto and Asahi wares (see pl. 15). For celadons and *temmoku* iron glazes, reoxidation at

1200 degrees instead of 1100 degrees centigrade is favored. Generally, the reduction phase should not be heavy.

To achieve late reduction, The kiln is fired in oxidation to 1220 degrees centigrade then in reduction to maturity. This schedule is effective for Oribe and Agano copper green glazes, for Shino ware, for opaque rice-straw ash glazes, and for green ash glazes (Seto, Iga).

Quick cooling, letting the kiln cool at least as fast as it heated, is ideal for transparent glazes, celadons, and *temmoku*. Slow cooling—that is, keeping the kiln going just enough to maintain the temperature above 1200 to 1250 degrees centigrade for several hours—is beneficial for Shino ware (oxidation during cooling), copper reds (reduction), and the iron glazes *irabo* (oxidation) and *soba* (oxidation or reduction).

SEVEN

Five Traditional Approaches

Even with access to reliable clays and forming methods, a large repertory of glazes, and efficient kilns, good results are never guaranteed. Technique is simply not synonymous with good potting or with a good pot. This is seen clearly enough in the fussiness and technical excess of the porcelains of late imperial China. Japan has also had its episodes of slavish imitation of foreign models and infatuation with garish shapes and decorations. Present-day Japanese potters, too, especially those seeking recognition in the country's all-too-numerous grand prix exhibitions, are not above meretricious effects. The overall experience of the past four centuries, however, suggests another approach, one in which the potter grasps the latent possibilities of the materials and uses techniques to enable these possibilities to emerge. In this sense, the potter acts more as a mediator than as a manipulator, helping clays, forms, decorations, glazes, and kiln effects to emerge and combine with the greatest effectiveness. This way of work emerges first, and most felicitously, in the ceramics of the late sixteenth and early seventeenth centuries. While these pots represent only a fraction of a long and very diverse Japanese ceramics history, they established a powerful precedent in terms of style, and even more important, basic approach, which inspired each succeeding generation in varying degrees.

BIZEN WARE

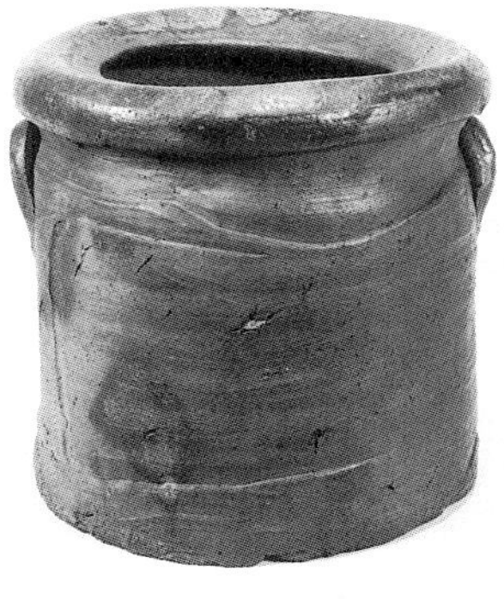
128. Bizen ware

Among the several unglazed stoneware traditions that survive from medieval times, Bizen ware (see pl. 2; fig. 128) fired from the twelfth century in and around Imbe village, Okayama Prefecture, amply demonstrates how potters evolved a forceful style from the simplest materials and methods. Since there is no use of applied glazes in Bizen ware, its life comes directly from clay and fire. From the late sixteenth century, Bizen potters began to use a fusible stoneware clay found one to three meters below the surface of rice paddies. This clay, although finely decomposed and plastic, was less than ideal. Possessing both a high percentage of iron (3 to 6 percent) and a high shrinkage rate, it was not amenable to glazing. Furthermore, the clay contained a lot of organic matter, and it would bloat and shatter if fired at the standard rate.

As a solution, the potters simply stacked the raw ware unprotected in the kiln and fired it once for a very long period. It was this open stacking and slow firing that led to many distinctive effects. The clay body developed an ironlike hardness and it took on a rich, reddish brown color, especially in the lower chambers of the kiln. Pieces shielded from direct flame by accumulations of ash assumed reduced surfaces of blue and gray. This is called "kiln change." Ash from the combusted pine fuel melted on the pots and created a spotty glaze effect known as "sesame seeds." When the spots ran together to form a glaze and rolled down the pot, the phenomenon was called "jewel droplets." On some wares stacked toward the rear of the kiln, a fine ash settled all over the pot and only partially melted, resulting in a melon-skin texture called "dry sesame seeds."

The potters were also quick to notice how particular kinds of stacking could lead to decorative effects. If a cup was placed upside down over a bottle, the shielded upper part of the bottle took on a different color and texture from its base, a finish called "upside down firing." When refractory clay pads, covered with rice chaff to prevent them from fusing to the adjacent surfaces, were placed on or wedged between pots, a round resist pattern developed (see plate 2). To keep stacked pots from sticking together in the kiln, some were wrapped with rice straw prior to loading (fig. 129). The alkali in the straw combined with the iron in the clay to produce red and brown streaks. This effect is called "fire cording," and it is now exaggerated by firing the wrapped pots in a saggar.

Other more recently developed techniques include salting the kiln at the end of the firing to produce a blue, sometimes glossy surface, and coating wares with a fine red clay slip and firing them in a saggar to produce a soft, reddish texture. Today, kiln change effects are exaggerated by inserting charcoal fragments and powdered charcoal into select areas of the kiln at the end of the firing. Applying

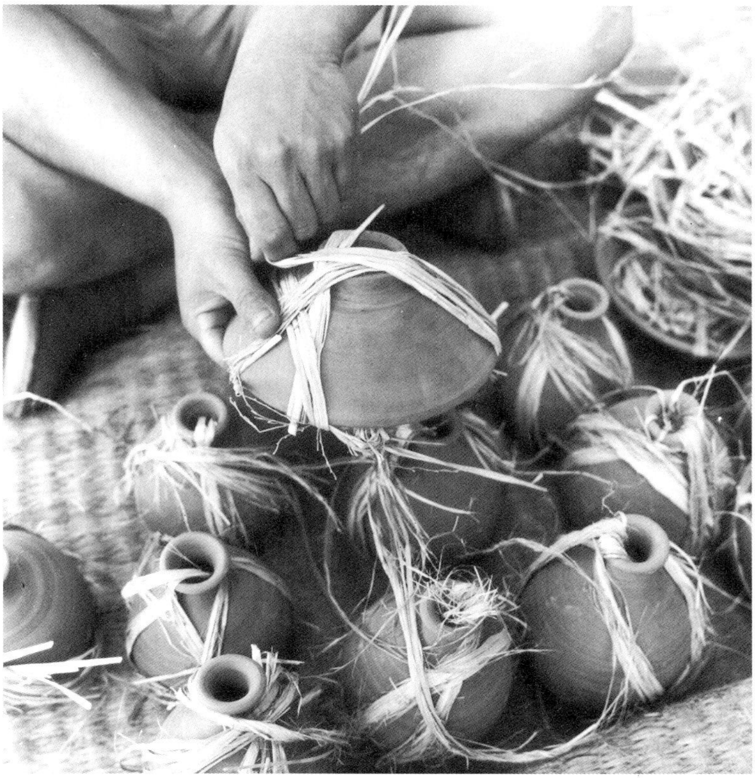

129. Preparing Bizen pots for "fire cording" effect

pine or chestnut ash to the pots prior to the firing replicates the sesame seeds effect. Over the past few decades, most Bizen workshops have enhanced the traditional effects with a multichambered climbing kiln designed to produce maximum flame circulation and ash deposition as well as a diversity of atmospheres. Key structural features include a very large firebox that doubles as a chamber in which pots are directly exposed to primary combustion; a small crossdraft chamber located between the firebox and the first chamber, which exposes pots stacked within to a fast, concentrated draft; and an absence of blowholes, which minimizes heating loss. This kiln has a tall chimney to produce the strong draft. Firing is extraordinarily long, requiring between seven and twelve days, and about 1200 bundles of split red pine are consumed in the stoking.

At first glance, the Bizen effects seem like gifts from the kiln gods, since the potter is excused from the responsibilities of decorating and glazing. Unfortunately, many Western potters who have reproduced Bizen or other wood-fired, unglazed stonewares have accepted such effects as serendipity, when in reality the authentic Bizen approach is quite exacting and intentional. Since Bizen clays rarely warp and there is no thick applied glaze, there is nothing to soften the profile of the pot. When fired, the ware retains much of the nakedness of a pot fresh off the wheel. It is therefore necessary to provide an uncommon degree of softness and sensitivity in the forming stages. The pots are not placed randomly in anticipation of random effects. Rather, past successes and failures teach the potters the peculiarities of each chamber, shelf, and spot, and these are taken into consideration in the stacking and firing. Consequently, Bizen potters maintain that good or bad stacking is tantamount to obtaining good or bad results, and since Bizen potters fire only once or twice a year, a lot is at stake. In short, Bizen technique is not simply the result of an arduous kiln firing and abundant ash deposition, but careful channeling of flame and fuel. Considering that most Bizen potters use only one clay body and one type of firing, the variety of effects is astounding.

From an outsider's point of view, Bizen may at first seem plain, dry, and somewhat cold due to its characteristically severe silhouette and unglazed surfaces. But with greater acquaintance, each pot tells of its experience in the kiln: where and how it was stacked, and how flame, ash, and carbon played on its surface. Reading this "autobiography" is in fact one of the most distinct pleasures of using Bizen ceramics.

MINO WARES

From the late sixteenth century, kilns in the Mino district centered around Tajimi and Toki villages, Gifu Prefecture, exploded with a series of creative products, particularly Shino, Yellow Seto, Black Seto, and Oribe. Recent excavations of merchant and warrior households in Kyoto, the national capital of the period, show that dealers bought kilnloads "as is" from Mino and had them shipped to the city. There merchants and tea masters looked over the wares and chose the ones they thought were interesting and useful. These preferences of course influenced the potters to make even more innovations. The creative latitude of this relationship, coupled with the long heritage of ceramics manufacture in the district, produced very distinctive pottery.

Shino Ware

Shino ware (fig. 130; see pls. 7–9) is famous for its thick, snowy-white glaze and distinctive red "scorch" marks, which seem to well up through the glaze. It has fascinated generations of Japanese and was singled out by novelist Kawabata Yasunari in his masterpiece *A Thousand Cranes*. The clay used for Shino is a partially decomposed secondary clay of the *gairome* type. It has little plasticity, so the forms tend to be thick, but the body is lightweight after firing and has a soft texture. This is probably the reason that it is called moxa clay—it possesses the soft fluffiness of moxibustion compound. Most important, the high alkaline content of this clay may have contributed to the distinctive surface effects.

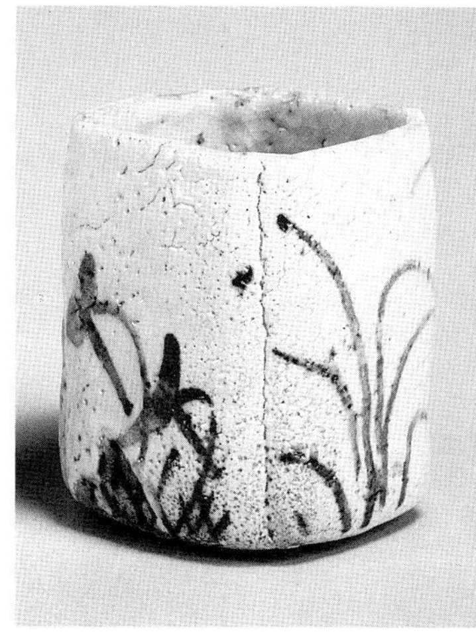

130. Shino ware

Typical Shino shapes include small and medium-sized dishes, teabowls, and other utensils for the tea ceremony. The squat, cylindrical shapes of Shino teabowls have drawn particular attention, due to their apparent lack of an antecedent. It has been suggested that in the course of making the low, cylindrical saggars for the ever-popular *temmoku* teabowls, potters observed that the saggars themselves had a broad, pleasing feel in the hands and designed tea bowls accordingly.

The abbreviated, whimsical patterns of Shino ware were painted in *oni ita*, mentioned in Chapter 4 as a local rock high in iron and manganese. For a reverse effect, some pots were covered with a slip made from this rock, and designs were scratched through into the white body in a sgraffito technique. In some pieces, the potter left one area unslipped and spontaneously filled out the design in sgraffito, based on the suggestion provided by that reserved white area. This is another good example of letting the materials and processes speak. Depending on their placement in the kiln, the slipped pieces fired either a rich red (oxidation) or a light gray (reduction).

Although it was believed that Shino glaze was composed of almost pure feldspar, recent studies have shown it to be an ordinary ash/feldspar glaze. The characteristic Shino coloration and texture was a result of the firing. The Shino kilns were semi-subterranean, very moist, and inefficient. During the prolonged firing, the glaze failed to melt completely and tended instead to either thicken or crawl on the surface. But the moist atmosphere and alkali in the clay did encourage the emergence of subtle red blushes. Some of the best contemporary Shino I have seen was produced in just such a circumstance, in a tunnel kiln fired during a rainstorm. The firebox actually flooded, and five days of firing failed to push the temperature much over 1200 degrees centigrade. Yet the Shino glazes were superb. Some contemporary Shino potters actually "irrigate" their fossil-fuel kilns with water pipes. Today Shino glazes are frequently simulated by adding a little bit of clay and salt (1 to 3 percent each) to feldspar (90 to 95 percent). In this mixture, chloride of sodium in the glaze probably combines with the iron in the body to produce the pinkish red colors. (Traditional Shino glazes have almost no sodium, except that which may have been provided by the clay.) For Shino, the kiln should be cooled slowly—about 30 degrees centigrade an hour in an oxidation atmosphere.

Yellow Seto Ware

Conventional opinion holds that Yellow Seto ware (see fig. 105) resulted from an unsuccessful attempt to make celadon. Possibly aware that the Chinese made simple celadon glazes with only ash and clay, Mino potters may have tried the same, yet from the looks of the devitrified yellow glaze, the kiln failed to melt the glaze and produce the reduction atmosphere necessary to create celadon. The potters may have added a ferrous clay to the Shino glaze to achieve the characteristic Yellow Seto color and texture. (Yellow Seto is also said to be a Shino glaze with natural iron in the feldspar or ash, with a bit of silica added.) The Mino potters must have considered the results a success, because they made modifications such as painted copper-green accents, which could only have been intended for oxidation. Interestingly, the bright color and dry texture of the best Yellow Seto glazes, sometimes called "deep-fried tofu," are effects that contemporary potters have failed to reproduce.

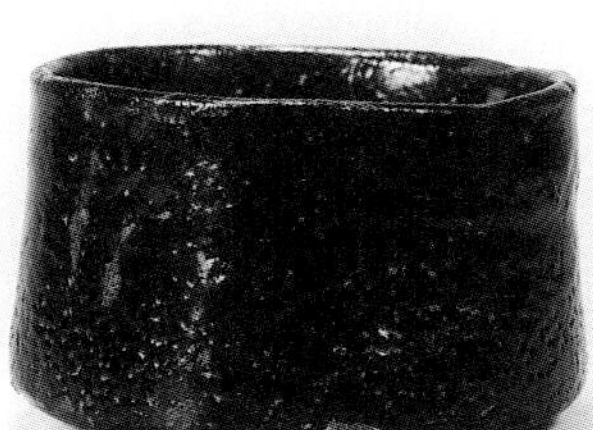

131. Black Seto ware

Black Seto Ware

Prior to the discovery of Black Seto ware (fig. 131), Japanese potters were thwarted in their attempts to achieve a true black glaze. The earliest iron glaze was produced by adding a clay containing a high percentage of iron to wood ash. The problem is that these early ashfluxed stoneware glazes could hold that much iron in solution only when the glaze was molten. In the cooling cycle, the iron crystallized, resulting in an opaque brown color.

During the sixteenth century, Mino potters gained access to metal tools and began to use tongs to withdraw test pieces from the kiln as it approached peak temperature. The potters must have realized that when they withdrew tests of their standard iron glaze, it was jet black, with the iron frozen in solution, instead of the usual brown. In some cases, these draw tests may have been conducted with regular pots, which would have suggested sudden removal as a standard procedure. In any case, simple black cylindrical tea bowls of the type now known as Black Seto were made in Mino kilns in the late sixteenth and early seventeenth centuries. Only a few pieces, mostly tea bowls, survive, and that may be connected to the limited number and sizes of pieces that one could pull through the spyhole of the single-chamber kiln then in use.

The transformation of chance occurrences into conventional style is part of virtually every ceramic tradition. In the case of Japanese ceramics, however, the *process* of discovery remains a part of the style. Although more convenient ways have been found to make black-glazed stoneware, the qualities of split-second judgment, as well as the drama of pulling a white-hot bowl from a roaring kiln, have endeared Black Seto to potters and users. This is indicated in another name for the ware, extraction black *(hikidashi guro)*, and in the fascination that users have for the marks that the tongs made on the molten glaze at the moment of withdrawal.

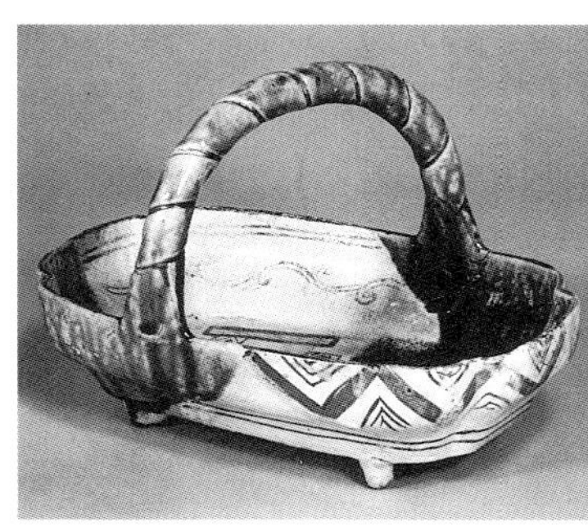

132. Oribe ware

Oribe Ware

Oribe ware (fig. 132; see pls. 10, 11) began to be fired around 1600, and although it used the same glaze as Shino ware, it was different in every other respect. Where Shino pots were formed largely on the wheel, many Oribe pieces were made with the drape-molding technique. In contrast to the limited repertory of shapes in Shino ware, Oribe ware shows a great variety, especially those based on fanciful combinations of squares, rectangles, and circles. Furthermore, Oribe-ware potters had access to the first multichambered, multivaulted climbing kiln built in Mino. This kiln fired evenly and quickly.

It was now possible to melt the Shino glaze completely into a transparent coating, and the potters immediately understood the new decorative possibilities. For example, red and white clay bodies were juxtaposed in drape-molded dishes, or white slip bounded with underglaze-iron outlines was used over a darker body. The transparent glaze made these contrasts emphatic. The potters also added dipped or poured accents of copper green glaze. This probably began as green painting, but when the potters realized how fluid and unpredictable the material was, they applied it amorphously. These green patches combined with the offbeat shapes, clay-body combinations, and exuberant iron painting to create one of the boldest and most diverting styles in Japanese (or any) ceramic history.

KARATSU WARE

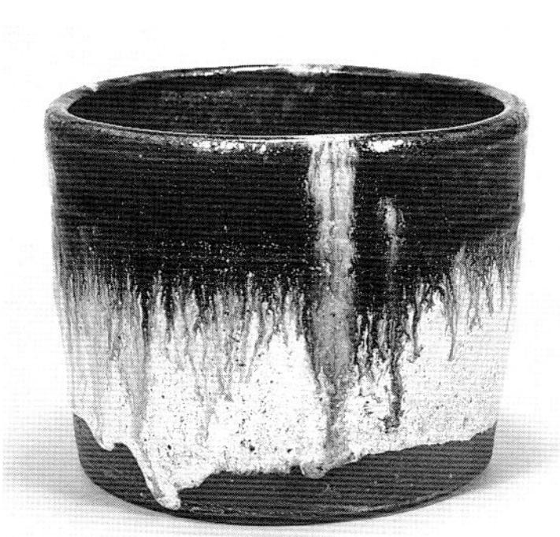

133. Karatsu ware

The term Karatsu is applied to a group of loosely-affiliated kilns that have operated in Saga and Nagasaki prefectures, Kyushu, from the late sixteenth century to the present. So productive were these kilns that the term "things from Karatsu" *(karatsumono)* became synonymous with ceramics in the western part of the country. Unfortunately, relatively few antique Karatsu (fig. 133; see pl. 4) wares are preserved, and those that survive tend to be regarded as masterpieces for use in the tea ceremony. This is at odds with the popular reputation of Karatsu as a major producer. In the past decade, however, the excavation of household sites in cities such as Kyoto, Osaka, and Kanazawa has made clear the full scope of Karatsu production: between about 1590 and 1640, the Karatsu potters absolutely dominated stoneware markets in western Japan. In order to sustain that high level of production, it was necessary to work efficiently, and in order to maintain demand, the ware had to be appealing. The Karatsu potters found a balance.

The typical clay is the partially decomposed secondary type, derived from decomposed granite. It is usually very plastic and contains a very high percentage of quartz particles. The granular texture is a hallmark of the style, contributing to the texture, fired lightness, and insulating qualities of the ware. Karatsu potters trim the clay while it is still very moist, and this makes the clay take on a soft, shredded texture, which in its most conspicuous form is called "crepe" (fig. 134). Interestingly, the Karatsu potters throw their pots in the conventional clockwise direction but frequently trim them counterclockwise, the direction in which the wheel is turned in Korea. This accentuates the crepe effect.

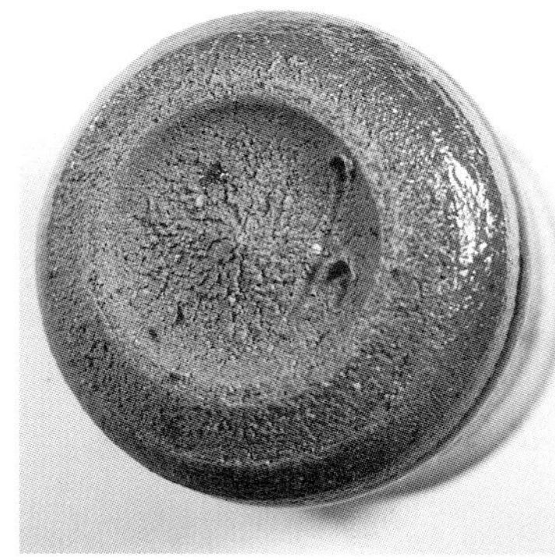

134. Crepe effect

Karatsu wares are made either by throwing or by coil-and-paddle combined with throwing. Judging from the nature of the forming marks on old pieces, the throwing was very quick. Pieces made for use as food vessels were frequently altered as soon as they could hold their shapes. There are countless variations, but they are all delightfully simple: edges are turned up or indented, or corners are formed. Rarely does one see the more contrived deformations typical of Mino ceramics of the same period. The larger coil-and-throw shapes at Karatsu are also simple and organic.

Decorative schemes relate closely to the clay body. When a light-colored clay was in use, the pots were left undecorated or painted in underglaze iron with a brush made by crushing one end of a bamboo-grass stem. Designs of reeds, pines, and plovers were favorite subjects, each consisting of just a few simple brushstrokes. A transparent glaze made of roughly equal amounts of feldspar and wood ash coated these pieces. Frequently the iron in the painting precipitated through the glaze in a pleasing "rust" effect. When dark clays were used, an opaque rice-straw ash was applied and broke handsomely over the darker body. Frequently an iron glaze was used together with the opaque glaze. Where the glazes overlapped, the interaction was lively, resulting in mottling and streaking.

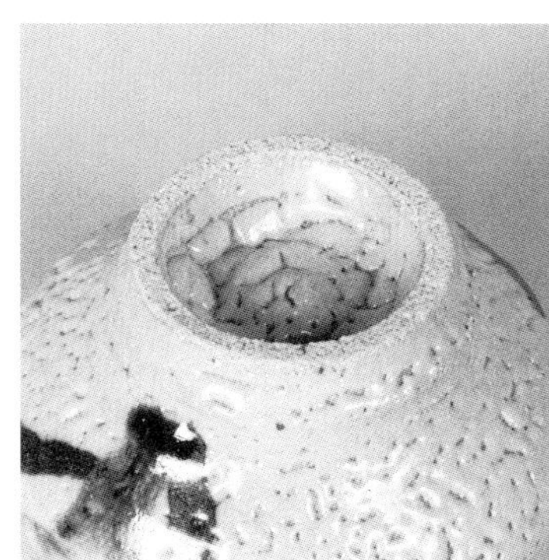

135. Crawl effect

Firing of the early Karatsu kilns was probably accomplished within a day, since the kiln was loosely packed, and the porous Karatsu clay easily withstood rapid temperature rises. But in a quick firing, there was much uneven heating. No saggars were used, and the bottoms of pots, especially those stacked one on top of another, frequently failed to mature and retained the pebbly or "crawl" texture characteristic of immature glazes (fig. 135). But users liked this effect, also found on the Korean ceramics that inspired Karatsu ware, and called it "sharkskin" after the pebbly skin of that fish. Coarsely trimmed surfaces enhanced this effect. The Karatsu stacking also shows how the potter adapted his products to the kiln. The standard transparent ash glazes, which melted at a comparatively

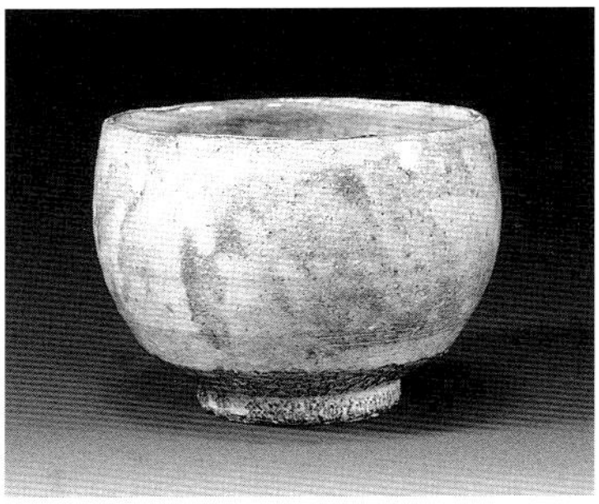

136. Hagi ware

low temperature, were packed in the cool spots of the kiln, usually the rear of the chambers. Iron and straw ash glazes, which were more refractory, were put in the hottest spots.

It is tempting to romanticize Karatsu based on its "exotic" Korean heritage, and too many critics have offered a kind of patronizing praise for its "naive" and "childlike" beauty. But can any potter who makes hundreds of wares daily for a mass market be considered naive? Karatsu speaks of a particular type of expertise born from long experience, where concentration and relaxation meet. Perhaps more than in any other Japanese ceramic, Karatsu shows a balance between discipline and freedom that is useful as a model for all studio potters.

HAGI WARE

Hagi ware (fig. 136; see pl. 6) is fired in and around Hagi, Yamaguchi Prefecture. Like many other wares in western Japan, it was first made at the beginning of the seventeenth century by potters taken from Korea during the two Japanese invasions of the peninsula in the 1590s. Distinctive to Hagi ware was its strong support from the local daimyo, who was especially interested in developing production of tea bowls in the Korean manner, such teabowls having gained enormous prestige nationwide since the late sixteenth century. As a result, Hagi potters developed only a few techniques, most dedicated to enriching the parent style. Judging from extant pieces and excavations, potters spent most of their time making tea bowls and small food dishes. Underglaze painting was not used, and almost all of the work was covered with a simple transparent glaze. Within these confines, the Hagi potters focused on clay, its texture, color, lightness, and interaction with glazes and fire.

Hagi potters use three raw clays. The most important one is Daidō clay, a light-gray colored secondary clay rich in quartz particles and very plastic. Mitake clay is a nonplastic, kaolinic clay, colored yellow-white. Mishima clay is a basaltic clay, neither plastic nor refractory, but rich in iron. These are variously blended, but the standard practice is to combine 70 to 80 percent Daidō clay with 20 to 30 percent Mitake clay. The resultant body is called "Korean" clay, and is famous for its subtle pink-orange hue, commonly called "loquat" because of its resemblance to the color of the fruit. Other standard bodies derive from additions to the Korean base: a coarse body called "devil" is made from a 10 percent addition of fine sand; a dark "Mishima" body is made from a 20 percent addition of the aforementioned Mishima clay.

These blends are sensitively combined with slip and glaze. For example, the standard Korean body is often coated with white slip. Knowing that dipping the pot in slip would completely obscure the subtle coloration of the clay, the potters instead chose to ladle on the slip, allowing it to irregularly veil rather than cloak the body. Contemporary potters sometimes blow on the wet slip to create thin areas. When a more stark contrast was desired, a thicker slip was applied to the iron-rich Mishima body. The iron in the body would burn through some areas of the slip to create a toasted appearance. The effect is especially conspicuous in reduction.

These combinations were sufficiently expressive to encourage the potters to rely almost entirely on transparent glazes. Mindful of the natural temperature variations in the kiln, the potters formulated clear glazes from 10 parts feldspar and 3, 5, or 8 parts ash, depending on placement. Wares covered by glazes with low percentages of ash were placed in hot spots, while those with high percentages of ash were placed in cool spots. That the flame was invited to play on the wares is evident in the absence of saggars. Wares were placed on round shelves stacked in columns. Interesting effects developed even in the cooler spots of the kiln, which would encourage the loquat-colored blushes through reoxidation and soft, pebbly glaze textures resembling the "sharkskin" seen in Karatsu. Over the course of the firing, quartz pebbles from the Daidō clay projected from the surface, often creating fissures in the slips and glazes. This is called "stone explosion." These fissures also invite moisture into the glaze in the course of use, and the subsequent discoloration is called the "raindrop" effect. Tea masters say that Hagi wares have seven stages of color transformation, underlining just how much the unaffected natural qualities of Hagi ware became its greatest asset.

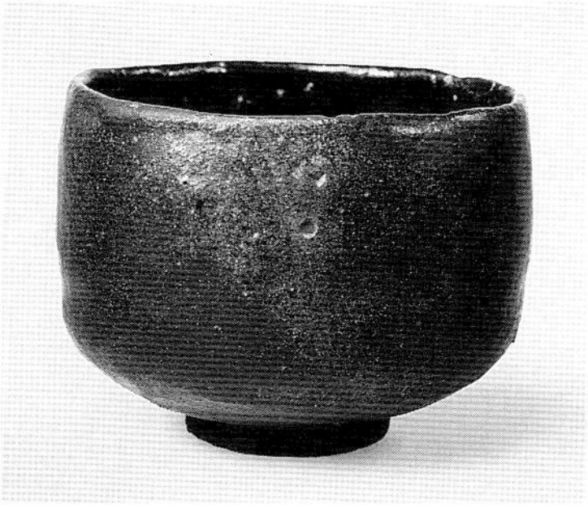

137. Raku ware

KYOTO WARES

Shortly after Kyoto became Japan's capital late in the eighth century, ceramics began to be manufactured in response to the demands of imperial and religious institutions. These objects were not unlike the products of other regions, and it was not until the late sixteenth century that a truly distinctive Kyoto style emerged. The catalyst was the tea ceremony. Kyoto ware differed from all the wares surveyed above due to the proximity of Kyoto potters to cultured patrons, who influenced the designs either directly or indirectly. At times excesses in taste led to mannerism, but the early Kyoto products, Raku ware and Kyoto stoneware, achieved an impressive balance of elemental power and refinement.

Raku Ware

Recent excavations of the houses of ceramics wholesalers in Kyoto have revealed not only stonewares from all over the country but lead-glazed earthenwares as well. The discovery of a few of the latter inside saggars suggests that earthenwares were made locally. Among these pieces are tea bowls of the austere, handmade variety known as Black Raku (fig. 137; see pl. 12). Where did this concept come from? Since both Raku and the aforementioned Black Seto feature withdrawal from the kiln at peak temperatures, it has long been debated which one was first. But since many of the wares excavated together with the Raku fragments are from Mino or are Mino derived, Black Seto may indeed have been the prototype.

In aspects other than the quick withdrawal from the kiln, however, the wares are different. Black Seto is formed and turned on the wheel and fired to stoneware temperatures. Raku is hand-carved, with a lower-firing earthenware glaze. Even though Raku bowls have a random, spontaneous look, their style probably derives from a conscious impulse by tea masters to tailor ceramics closely to tea drinking. There were few potters working in Kyoto at the time, so tile makers evidently were recruited to fashion the ware. These craftspeople were experienced in sculpting ornamental tiles, and that is the approach they took in forming the bowls, which were made by a combination of pinching and carving. The sculpted

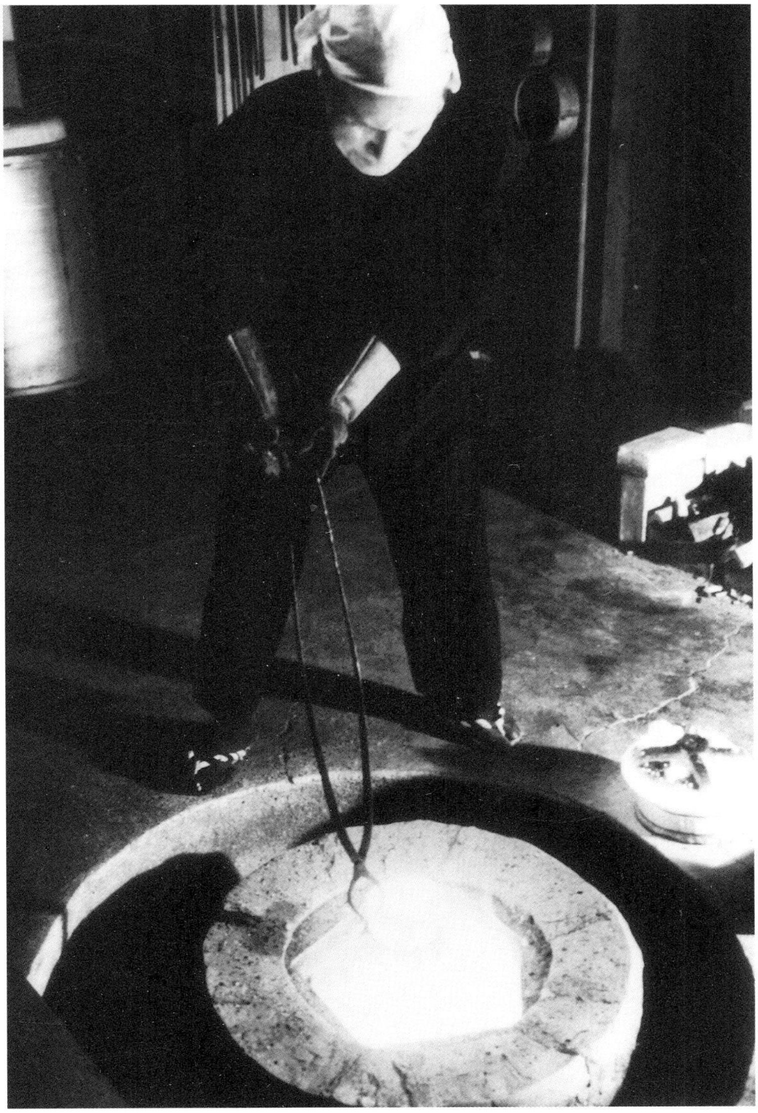

138. Ogawa Chōraku firing Black Raku ware

forms, relieved of the hard geometry of the potter's wheel, proved to be very pleasing to hold. As for the glazing, the Black Seto method probably inspired the potters, but they had to duplicate the style with their own earthenware technology. To make a black glaze, they mixed lead with Kamogawa stone, a ferrous rock from the bed of the Kamo River in Kyoto. This glaze was applied by painting or dipping.

Although the first Raku kilns were probably the standard cylindrical updraft kilns used for firing other earthenwares, within a few decades a distinctive method of Black Raku firing evolved. The Black Raku bowls were fired individually in a lidded saggar set inside a

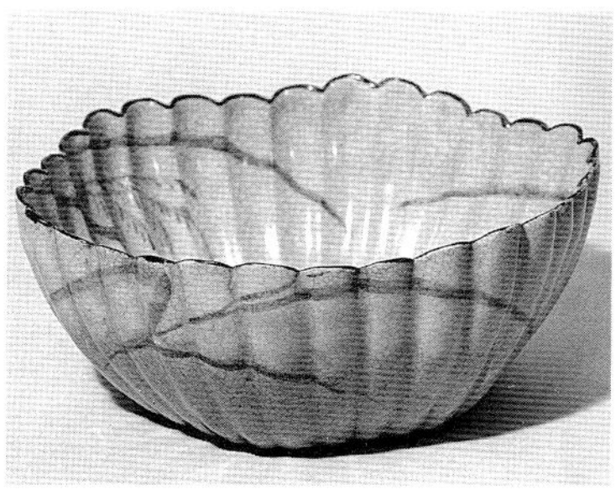

139. Kyoto stoneware

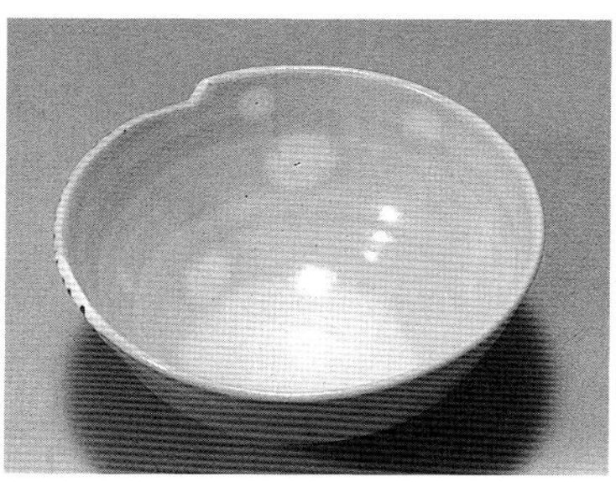

140. Halo effect

cylindrical kiln. A very hard charcoal cut into small sections was stuffed into the space between the saggar and kiln walls. A bellows (today a blower is used) provided air to the charcoal through the bottom of the kiln. The charcoal was ignited and brought up to temperature, about 1100 degrees centigrade for Black Raku. A bowl preheated at kilnside was inserted into the saggar, the saggar was closed, and the bellows was started. Charcoal was continually stuffed down into the combustion area, each time resulting in a shower of sparks and flames. Once the kiln was up to temperature, each firing took about five minutes. When the glaze was melted, as determined by stopping the forced air and looking into the saggar, the red-hot bowl was removed (fig. 138). It was placed on a clean shelf and allowed to air cool. In the Black Raku firings that I attended, water-quenching was never used. Post-fire reduction is a Western innovation.

When the Black Raku glaze was withdrawn from the kiln at peak temperature, it turned a semi-matte black color with a soft pebbly texture, which Raku potters call "citron skin." While not technically perfect, this made a good match with the irregular form of the bowl. Furthermore, the clay remained porous and unvitrified, with a low thermal conductivity that was ideal for a hot-beverage vessel. Raku is so highly regarded in Japan that it is easy to overlook these few simple premises.

Kyoto Stoneware

The clay used for Kyoto stoneware (fig. 139; see pl. 15), sometimes called Kurodani after the place in eastern Kyoto where it was mined, came from a streak of decomposed granite that ran underneath the city. This clay was a finely textured, very plastic *gairome* with a low percentage of iron. It was particularly well suited for painted decoration, which became a Kyoto specialty. In addition, the local transparent glaze cooled with a fine crazing pattern, imparting an additional softness to the ware.

The relatively low firing temperature, about 1230 degrees centigrade, and the thinness of application of the Kyoto transparent glazes occasionally encouraged the formation of pinholes, which occurred over larger grains in the clay. A halo of contrasting color formed around the pinholes if the kiln atmosphere changed in the later part of the firing (fig. 140). Seeing some decorative potential in this phenomenon, Kyoto potters added between 5 to 30 percent coarse fireclay to the Kyoto clay to encourage the pinholes. They also found that an application of a kaolinic white slip would enhance the color of the halos. Today this effect is produced in a mild reduction fire between 900 and 1150 degrees centigrade, followed by oxidation until 1230 degrees centigrade. The kiln is cooled slowly to prevent the glaze from turning greenish and the halos from disappearing. This technique is also employed at Hagi and at Asahi, a kiln just south of Kyoto.

BEYOND TECHNIQUE

There is sufficient consistency of approach in these traditions to suggest some underlying assumptions. These point to a way that uses technique to go beyond technique.

Clay: In the development of clay bodies, texture, color, and responsiveness take precedence over the conventional emphasis on smoothness, whiteness, and refractoriness.

Forming: Techniques of forming respect the clay's properties rather than coercing them.

Decorating: This stage is typically quick and minimal, retaining a ceramic quality. That is, plastic decoration echoes the form rather than distracting from it, painting is suggestive rather than representational, and decoration is kiln-sensitive, so that slips change color, and pigments develop metallic flashes.

Glazes: They are simple in formulation. Smoothness, gloss, and uniformity are important, but so is responsiveness to clay body and firing. Glazes that would be considered defective from a purist's point of view are admired for what they are.

Firing: Fire is an active agent in decoration, a notion that affects the construction of the kiln, which is not always efficient in the Western sense, and its firing schedule. The idiosyncrasies of the kiln are exploited rather than eliminated.

These characteristics serve to underline the uncommon intimacy of the Japanese potter with materials and methods, and the way in which the finished pots have exposed and further encouraged that kinship. Ideally, the relationship should be a dynamic one, responsive to change as resources become exhausted, environmental changes affect materials, and new generations of users demand new things. Unfortunately, the traditions surveyed above attained such great esteem that they were elevated into classics. Potters of other times and places tried to copy them, and in so doing, violated their own local materials, procedures, and consumer demands. As so often happens in the transmission of art, style was reduced to a formula—a certain shape, a certain design, a certain glaze—rather than a spontaneous manifestation of human activity in response to a given environment. A similar misunderstanding appears in the Western reception of Japanese ceramics, the subject of the concluding chapter.

EIGHT

Japanese Methods and Western Potters

The transformation of clays, stones, and ashes into functional containers is the basis for all traditional ceramics techniques. But tradition itself is much more: it is a concept that connects the technical and functional components of the art to the culture as a whole. When a certain vessel shape or glaze is identified as traditional, it acquires, in effect, a history. It stands not only for itself, but for all its precursors. In the Japanese art world, the invocation of tradition is especially weighty, because traditional style is thought to embody not only the national past, but virtues central to Japanese identity such as love of nature, social cohesiveness, and dedication to work.

An unglazed flower container from the Shigaraki kilns (fig. 141) exemplifies how these factors work in ceramics. For the knowing user, the shape of the pot may prompt the arrangement of certain seasonal flowers. These in turn can evoke the particular emotions that people feel at that time of year. A pattern of accidental ash deposition may suggest a landscape for some, while for others, the fragments of quartz and feldspar that protrude from the clay body may recall the decomposed granite hills of the region around the kiln center. The rather dull, unglazed surfaces of the vessel may provide a reminder of the poetic "chilled and withered" images that the great Japanese tea masters found in these pots. The object may summon memories of people who used Shigaraki ware, ranging from past personages to personal friends. The artless anonymity of the vessel may also produce an identification with rural laborers, with whom most urban Japanese feel a nostalgic, ancestral bond. In short, a traditional pot is much more than a functional container, for its purchaser buys into a historical and aesthetic universe that continues to expand with each new deployment. In this sense, use and cultural identity become synonymous.

Cultural associations are certainly not unique to Japanese ceramics, but in Japan they have an exceptionally high level of complexity, specificity, and social reinforcement. The latter is guaranteed by a large and prosperous traditional art world that includes

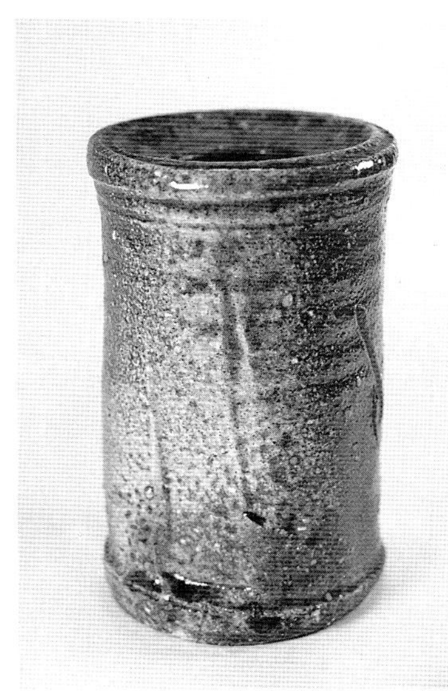

141. Shigaraki-ware vase

publishing houses, galleries, and museums, as well as the practitioners of the tea ceremony, flower arranging, and traditional cuisine. Enthusiasm is widespread, and it is not uncommon to see men and women of all ages browsing in exhibitions and buying pots in stores, small galleries, and at potter's shops. The prestige that tradition enjoys in public and private life in Japan also encourages new generations of potters to engage in traditional ways of work. Some of these features were surveyed in the first chapter.

But if tradition is culture-specific, what happens when there is an impulse to work in, or in some way draw from, the Japanese tradition in a non-Japanese environment? This closing chapter will take up the issue of how potters outside Japan have treated the Japanese legacy. What is central to a Japanese pot? Is it the way it looks, the nature of its materials, or how it functions? Or is there an intangible "way" that goes beyond the surface? Moreover, how can this Japaneseness be reconciled with the economic, social, and aesthetic concerns of contemporary Western potters? Is it possible to establish a critical standard in this field, or are all such attempts simply rhetoric, a means of justifying one's own particular interests?

MODES OF TRANSMISSION

The West's historic encounters with Japanese ceramics provide some lessons about the difficulty of an authentic artistic transmission between cultures. Borrowing from a foreign culture never takes place within a vacuum. Influence has at least two different dynamics: one culture gives, and another receives. The Western reception of Japanese ceramics, then, has been fashioned by what the Japanese have made available and what the West has been willing or able to accommodate. This is further complicated by the fact that both techniques and aesthetics were transmitted, but in different ways and at different times. When these strands are untangled, however, they reveal how the West has variously learned, invented, or, for that matter, ignored aspects of Japanese ceramics.

The earliest episode of influence, covering a period between the mid-seventeenth century and the end of the nineteenth century, is marked by exoticism and fantasy. Since there was little travel between Japan and the West, the source material was largely confined to the pots themselves, gilt and enameled export porcelains from the Arita kilns in Kyushu and later from kilns in Kyoto, Kutani, and Satsuma. These objects were usually made in shapes and decorations thought by the Japanese to represent foreign taste. The importers, for their part, believed they were getting something deliciously alien, and appreciation of these objects was part of a well-developed appetite for the ornate and foreign that characterized the Baroque and Rococo eras in Europe. Specifically, consumers enjoyed "quaint" or "charming" symbols such as pagodas, folding fans, or kimono-clad females, and just such a vocabulary appears on the imitation porcelains produced in Europe after the mid-eighteenth century. Reducing a foreign culture to the level of entertainment, however, trivializes that culture and deprives it of its own voice. Without voice, there is no conversation, no way for the other culture to explain itself. Entire worlds become simple fads or fashions, and restless audiences move on.

The conclusion of trade agreements between Japan and Western nations in the 1850s facilitated far greater access to Japanese culture and material goods. An interest in what might be called the formal aspects of Japanese ceramics, technique and style, augmented and occasionally replaced simple exoticism. The Philadelphia Centennial Exposition of 1876 and the Paris Exposition Universelle of 1878 marked the first displays of historical examples of Japanese ceramics in the West. At the same time, reports were coming in from field observers like scientist Edward S. Morse, who emphasized the way in which the Japanese themselves used and admired pots. Morse also criticized the inauthenticity of the export style. Traditional Japanese sensitivity to materials and design also attracted the attention of proponents of the Western Arts and Crafts Movement, who were looking for alternatives to unattractive and alienating machine production. By the end of the nineteenth century, individual French potters such as Jean Carriès and Georges Hoentschel began to imitate simple forms and rustic surface effects in their own works. In England, the salt-glazed stoneware of the Martin brothers showed some influence from Japan.

These trends, however, were hardly mainstream. The ornate neo-Rococo or Victorian style continued in high fashion, and the newly introduced Japanese domestic styles, especially quiet, monochrome types, had more than a few detractors. Japan authorities Frank Brinkley and James Lord Bowes led the charge, alleging that Japanese tea masters had rescued a bunch of bland and deformed pots otherwise consigned to the "dustbin of history." Little did Brinkley and Bowes realize that the glittery overglaze enamel wares they held up as classically Japanese were nothing more than recently made export wares. And even though the contemporary audiences worldwide have been exposed to the quiet subtleties of native Japanese products, the reception has not always been enthusiastic. The nineteenth-century bourgeois sensibility, which equates value with quantity and explicitness of ornament, continues as a dominant influence in contemporary Western ceramics. Unfortunately, all the surface information obscures and devalues other, less obtrusive messages in clay.

Bernard Leach, whose career spanned much of this century, is the seminal figure in this formalist era, for he assembled disparate elements of Japanese practice into a comprehensible system, one that potters might not only admire, but apply in their own work. Leach also represents both the romantic anti-industrialism of the late 1800s and the more intimate experience of Japan enjoyed by twentieth-century potters. His art education in England brought him into contact with the Pre-Raphaelite and Arts and Crafts movements, which attempted to reassert heart and hand over excessive reason and technology. In addition, Leach's experience as a potter and observer in the field brought the Japanese ceramics world into closer focus. He first began potting seriously in Japan in 1917 and worked there intermittently until 1920. He returned for brief periods of production in 1934 and 1952. In his native England, Leach teamed up with visiting potters Hamada Shōji and Matsubayashi Tsuneyoshi to build the first Asian-style climbing kiln in the West; Japanese techniques abound in his *A Potter's Book,* first published in 1940. Unlike the earlier porcelain producers and Arts and Crafts adherents such as Carriès, Leach also had a spiritual vision: appreciation of Asian ceramics was not only of artistic interest, but a step toward an emerging global culture, where East and West would meet. While Leach never underwent a traditional apprenticeship in Japan, and while he frequently criticized Japanese aesthetics, siding with his friend Yanagi Sōetsu against the conscious distortion so abundant in later tea-ceremony wares, his extensive writing and lecture tours influenced many Western potters either to travel to Japan or at least to explore some facet of the tradition.

Artistic influences are not limited to the visual, tangible, or technical. A third period of Japanese influence on Western ceramics, one in which the spiritual and processual aspects of the craft came to the fore, commenced with the end of the Second World War. Agendas on both sides of the Pacific colored this era. For the Japanese, their own traditional culture offered refuge from the stark reality of wartime defeat. It also served as a way of self-representation untainted by the recent militaristic and imperialistic past. This image was also attractive to Westerners, who gradually abandoned their negative stereotypes of Japan for a more benign one: Japan as a repository of spiritual values and artistic treasures. Due to this new enthusiasm and their position as conquerors and occupiers, Westerners gained unprecedented access to pots and information, and all kinds of Japanese ceramics began to be exhibited, collected, published, and even imitated abroad. The booming peacetime economy in the United States made it possible for more students, hobbyists, and professionals to enter the field. The counterculture movements of the 1960s and 1970s also lent attraction to handmade ceramics as part of an alternative, back-to-nature lifestyle. Japanese styles and techniques, ranging from wood fired stoneware to Raku, came into vogue.

While many potters continued to be interested in technical information, part of the appeal of Japanese ceramics stemmed from the expanded influence of Asian philosophy in postwar arts and letters. For example, writings on Zen Buddhism, especially those of D.T. Suzuki and Alan Watts, influenced a host of artists, among them John Cage in music, Gary Snyder in poetry, and Robert Morris in sculpture. None of their works read as Japanese in terms of style. Instead, this influence encouraged new artistic attitudes such as faith in change and randomness, elevation of the ordinary and ephemeral, and fusion of process and product. This seductive mindset, which seemed to align ancient Asian traditions with the restless and iconoclastic American way of thinking, attracted potters as well.

Japanese ceramics themselves played a role in this attraction. The imperfect shapes, casual decorations, and serendipitous firing effects of folk-art ceramics appeared to manifest Asian philosophy and contemporary Western fine art simultaneously. The drip glazing on Shōdai ware, for example, could be likened to the action painting of Jackson Pollack; the contrasting glaze panels on Ushinoto ware were comparable to the color fields of Mark Rothko; and distorted, kiln-ravaged Iga ware was a low-tech analogue to John Chamberlain's crushed auto parts. Like the contemporary objects, these pots seemed to express a single, momentary impulse, that of the intuitively gifted "fine" artist. The connections between Japanese practice and the new gestalt were especially nurtured in the 1952 American tour of Bernard Leach, Hamada Shōji, and Yanagi Sōetsu, in which the methods of the "traditional" Japanese potter Hamada were set against a background of Buddhist-inspired intuition and acceptance as expressed by Yanagi. For a new generation of American potters, here was a "way" of Japanese ceramics that was related less to form, technique, or function than to creative attitude and process. Thus, for potters such as Peter Voulkos or Paul Soldner, who moved vigorously fabricated, minimally decorated forms in the direction of nonrepresentational sculpture, Japan was, if anything, a reinforcement of their concern with gesture and spontaneous play. Potters working with traditional vessels also shared this dynamic. This new sense of Japan contributed to a renegotiation of the meaning of pot making itself, one revealed in the semantic drift of the postwar period: from "workshop" to "studio," from "skill" to "creativity," and from "craftsman" to "artist."

ALTERNATIVES

If the postwar Western infatuation with Japan was stimulating, it was also arbitrary and selective. As happens with most artistic borrowings, those elements that validated local practice were seized upon. The irregular shapes, kiln accidents, and asymmetrical glaze splashes on Japanese wares seemed congenial to the growing Western conviction that in art the process was just as important as the product or that the two were indivisible. In terms of ceramics, this also fitted well with the antimachine ideology of the Arts and Crafts movement, which resurfaced in the postwar crafts boom. Unfortunately, the antimachine canon reached extreme proportions in postwar studios and classrooms, where, for a time, every pot had to have gritty clay, prominent finger grooves, and iron-flecked reduction glazes. The more handmade a pot looked, the better.

Another selective Western focus was the attention given to the status of the potter. Since Japanese potters were and still are sometimes idealized as intuitively gifted, if not enlightened masters, it was relatively easy for Westerners to assume that attitude was more important for a good potter than serious manual training. This was also supported by an Arts and Crafts assumption: Bernard Leach and other craft revivalists stressed that given the general decline of the crafts in a world of industrial domination, only "artist-potters," individuals blessed with leadership and insight, could save them from extinction. In a famous rejoinder, Yanagi Sōetsu defended Hamada's heavy pottery as "better than being too light." Yanagi was suggesting that earthy ceramics had a sincerity that outdistanced any physical limitation, but such statements became food for misunderstanding when taken out of context. In Western studios the antimachine and expressive elements combined with "Japaneseness" to encourage and legitimize a lot of heavy, rusticated (not rustic) stoneware. Potter Harry Davis, a one-time Leach disciple who was able to distance himself from this conceptual baggage, called it "the odd cult of unskill." Unfortunately, most artist-potters were unfazed by such criticism. In adopting the mantle of the artist, they also developed an indifference to their ostensible audience. As in painting, the function of the pot was to transmit a personal message, and the interpretation of that message was the problem of the viewer. And if people were willing to buy Japan-inspired pots without subjecting them to the test of kitchen or table, so much the better.

This game could continue as long as the Japanese way was seen as new, exotic, or personally expressive. Now those days are over.

Japanese-style pottery is no longer considered avant-garde, and the cult of unskill has alienated many potential patrons. Indeed, although the root causes may be somewhat different, the same careless design and shoddy workmanship that in recent decades characterized American industrial manufacture has come to permeate functional ceramics. But reform is no easy matter for our potters. People believe that they need cars, but how many believe that they need handmade pots? Yet pots that combine an appreciation of natural materials, sharp craftsmanship, and a generous sense of utility can only make more believers, and in this regard the Japanese tradition provides an attractive and, moreover, adaptable approach. Here are some suggestions about how one might access and implement those assets without simply becoming another Japanese-style potter.

At the highest level of aspiration, it would be best to simply go to Japan and work. It is, after all, the only way to experience the craft and the culture on their own terms. There are many ways to pursue ceramics in Japan. Some people obtain easy employment as an English teacher, visit kilns during their spare time, and make pots at a neighborhood culture center. Others formally attend art school—there are major ones in Tokyo, Kyoto, Kanazawa, and Osaka. The nature of instruction at these schools is theoretical and experimental, much like that in Western institutions. Most people agree, however, that the way to get ceramics really into the bones, learning not only facts but a rhythm and way of clay work unique to Japan, is to become an apprentice. Finding a teacher is no easy task, since master potters consider foreign apprentices not so much assistants as responsibilities or even burdens. An introduction is the best way to get in the door. Some veterans suggest that even without such an entrée, the aspiring student can visit one of the larger pottery centers like Bizen or Mashiko to meet other Western potters already at work and consult with them about someone to work with.

The apprenticeship is rigorous and serious. The master assumes that the student intends to learn every step of the potter's work and therefore has the student start at the beginning. There is an initial year or so of workshop chores that include pugging, wedging, and recycling clay, mixing glazes, and cleaning up the workshop and kiln. No new work is introduced until these routines are mastered. This may test the patience of a visiting potter with limited time in the field, but it is the best opportunity to observe techniques and working rhythms. Around the second year, one begins making the smallest, simplest object in the production line, usually teacups. Each shape must be perfected before moving on to the next level. One former Japan apprentice mentions that her teacher expected one hundred identical pieces before allowing her to work on a new item.

In addition to what one learns, *how* one learns is important. The Japanese pedagogical style varies significantly from the Western model. Very rarely will a teacher directly tell the pupil what to do, for the gateway to learning a craft in Japan is not intellectual but largely nonverbal, physical, and intuitive. In a way this process is more direct than swallowing abstractions and trying to reproduce them consciously. The body learns through absorption and repetition, although the Westerner may become frustrated by the lack of verbal feedback. The student begins to feel isolated and confused, but with experience he or she will see the cues in the materials, procedures, and larger workshop rhythms. As the student begins to fit into the dynamics of the workshop and realizes how much is at stake in every move, rules that once seemed restrictive and arbitrary reappear as common sense. Gradually, a strong awareness dawns that *this* is what it means to be a practicing potter, not just a hobbyist or artist. At that stage the student realizes that there is something counterproductive in the Western tendency to think that personal expression and skill are opposites. Good, sharp technical skills are essential for a healthy functional potting tradition. No amount of conceptual tomfoolery can disguise a poorly made pot.

The Japan experience also offers benefits outside the workshop. As taciturn as Japanese potters tend to be, there are many opportunities to engage in shop talk. Chances abound to visit exhibitions, antique shops, flea markets, and restaurants. One former Japan student mentioned that experiencing pots in context helped her to understand that vessels were conceived with food in them—that is, in use rather than empty. But there is a fine line between stimulation and overload. Japan is a cultural smorgasbord that may leave the visiting potter with indigestion. It is important to sift through all the influences carefully and settle on, as one veteran says, "the kind of pots that are right for you."

Most apprentices return home. Here, in a sense, this book returns to its very beginning, that imaginary morning when one awakens to the day's work. The experiences in Japan with different tools, materials, instructional styles, workshop dynamics, and especially an environment so sympathetic to handcrafted ceramics may remain in the mind's eye and sustain the potter's dreams. But the work is in the West, and for that reason, and for those potters who cannot make the long journey, I offer some final thoughts. What are the enduring lessons of Japan, and how can they be reconciled with the concerns of contemporary Western potters?

First, the Japanese way suggests the value of fully integrating materials and methods into one's work. To put it another way, it is important to observe and develop the latent but very distinctive potentials of each material and method and harmonize those elements in the work. Typically, Westerners tend to take a tool kit approach to foreign influence, reducing whole traditions to discreet data bits that are subsequently compounded into a new style. There is nothing unusual or disingenuous about this practice: humans, by nature, break down and reassemble information. I contend, however, that just as traditional vessel making is a holistic art, a holistic understanding of Japanese ceramics offers more to the potter. That entails thinking not so much about this particular shape or that special glaze formula, but the process in which they originated. This may be accomplished by trying to reproduce Japanese clays, shapes, decorations, glazes, and firing techniques. But replication should not be the final goal. For example, many of the greatest Japanese styles originated in attempts to imitate Chinese and Korean models. What gave the results life, however, was not the rigor of the copying but a basic attitude: materials and methods are to be cooperated with rather than coerced, and unexpected results are not only delightful but suggestive of new style. This blend of pragmatism and humility is more worthy of emulation than any clay or glaze formula. The ongoing effort to develop an aesthetic in the process of *making* rather than using techniques as the means to achieve an abstractly conceived end will lead the potter closest to that hard-earned and pleasurable stage of integration that most people desire from work.

It is also important to recognize that the strength of Japanese ceramics lies just as much in consumption as it does production. At the beginning of this chapter, I outlined the power of tradition in Japan. Traditional Japanese ceramics were conceived in a system that embraced time, place, pot, and person, so that pots were not seen as isolated objects. One challenge to Western potters and their audiences is to explore the nuances of function, not just in the making, but as a part of the aesthetic, social life of the pot. Although traditional craftspeople had little need to contemplate such issues, today the maker can and must enter this world consciously. This means mastering the supply *and* demand side of ceramics, the latter through regularly using pots, washing them, and looking at them full and empty. Potters who cook have an advantage over those who don't. Eating and drinking are sensory experiences, and vessels are an important part of gustatory life. How does a particular clay body feel in the hands? How does it feel against the lips? How does the fired clay surface or the glaze sound when grazed by a metal fork? By a wooden spoon? What is the relationship between the way the vessel looks and how it feels when it is lifted? Does the decoration enhance or detract from these simple sensations? What kinds of personal and collective memories will the shape and decoration evoke? How does the pot fit the day's mood, the weather, the guests?

This way of thinking may evoke a domestic life instead of an artistic one. The concept of domesticity, charged as it is in the West with class and gender prejudices, may seem unattractive to potters and critics. But the Japanese have made kitchen and table into a critical arena, one where taste—in the immediate, visceral, nonelitist sense of the word—bestows the honors. Few would deny that their pottery is better for it. The synthesis rather than the opposition of art and utility is the very marrow of Japanese ceramics, and an active embrace of that relationship still remains the major frontier for Western functional potters. It is of course hard to ask such a breakthrough from potters alone. If only people treated each moment of their lives with equal preciousness, all pottery would become unimaginably rich.

At its very best, the Japanese legacy means more than using certain materials such as raw clay and wood ash. It does not necessarily mean making tea ceremony bowls, or living in a country house and firing with wood. It suggests, rather, renouncing artistic individualism, including the superficial distinction of working in a foreign style, which brought many Westerners to Japanese ceramics in the first place, and rededicating oneself to personal discipline and harmony with materials, processes, and the user community. It implies thinking about the pots as part of an ongoing process, a relaxed conversation with the world, instead of as a kind of personal decree. And if, at this point, "Japan" disappears, so will the unnecessary conceptual and expressive baggage of the artist-potter. Clay vessels will remain, ones that are not only pleasant to use but point quietly to a broader way of life.

Works Consulted

Cardew, Michael. *Pioneer Pottery.* New York: St. Martin's Press, 1969. Michael Cardew, a prominent disciple of Bernard Leach, never studied in Japan, but this elemental and eloquent approach to potter's materials and techniques transcends cultural boundaries.

Eguchi Akira. *Tōgei Nyūmon* [Introduction to the ceramic arts]. Osaka: Bunken Shuppan, 1987. Best of the numerous introductory guides to potmaking published in Japan.

Kanzaki Noritake. *Kurashi no Naka no Yakimono* [Ceramics in life]. Tokyo: Gyōsei, 1986. An illustrated exposition of the traditional uses of ceramics in Japan.

Kingery, David W., and Pamela V. Vandiver. *Ceramic Masterpieces.* New York: The Free Press, 1986. A sophisticated and lucid technical analysis of ten ceramic masterpieces, ranging from appearance and texture to microstructure. No Japanese ceramics mentioned, but inferences can be made.

Kōgei Shuppan, eds. *Kama kara Mita Yakimono* [Ceramics as seen from the kiln]. Eight essays by potters from different regions on building, loading, and firing kilns. Tokyo: Kōgei Shuppan, 1987.

Leach, Bernard. *A Potter's Book.* London: Transatlantic Arts, 1970. The classic.

Nihon Hōsō Kyōkai, eds. *Tōgei Nyūmon* [An Introduction to the ceramic arts]. Tokyo: Ryosō Kyōkai, 1987. A well illustrated guide for beginners.

Ōkura Enpei. *Yakimono Nyūmon no Subete* [The complete introduction to ceramics]. Tokyo: Nihon Bungeisha, 1974. The most complete of all the Japanese handbooks, unfortunately with poor illustrations.

Olsen, Frederick L. *The Kiln Book.* Bassett, California: Keramos Books, 1973. The best English-language description of Japanese kilns.

Ōnishi Masatarō. *Tōgei no Yūyaku* [Ceramic glazes]. Tokyo: Rikōgakusha, 1981. Ōnishi is widely regarded as a top authority on glazes for studio ceramics.

_____. *Tōgei no Dentō Gihō* [Traditional ceramics techniques]. Tokyo: Rikōgakusha, 1982. Ônishi applies his expertise to traditional techniques.

_____. *Tōgei no Tsuchi to Kama Yaki* [Clay and kiln firing in ceramics]. Tokyo: Rikōgakusha, 1983. The section on clay is interesting.

Philip, Leila. *The Road Through Miyama.* New York: Random House, 1989. Thoughts on contemporary Japan from multiple perspectives: foreigner, female, and apprentice potter.

Rawson, Phillip. *Ceramics.* London: Oxford University Press, 1982. A concise, authoritative overview of world ceramics by a historian-critic.

Sakai Manabu. *Yakimono Nyūmon* [An introduction to ceramics]. Tokyo: Gotō Shoin, 1986. Another potter's handbook, not as comprehensive as Eguchi, but with a few different techniques.

Sanders, Herbert. *The World of Japanese Ceramics.* Tokyo, New York, and San Francisco: Kōdansha International, 1968. This book opened up the postwar generation of readers to Japanese pottery techniques. Focuses on the work of the potters central to the postwar ceramic boom.

Satō Masahiko. *Yakimono Nyūmon* [An introduction to ceramics]. Tokyo: Heibonsha, 1983. Combines basic potter's techniques with historic masterpieces of East Asia.

Simpson, Penny. *A Japanese Potter's Handbook.* Tokyo, New York, and San Francisco: Kōdansha International, 1979. A handy English-Japanese guide for Western potters studying in Japan.

Tagai Hideo. *Nyūmon Yakimono no Kagaku* [An introduction to ceramic science]. Tokyo: Kyōritsu Shuppan, 1974.

_____. *Nyūmon Yakimono no Tekuniku to Kanshō* [An introduction to ceramics technique and appreciation]. Tokyo: Kyōritsu Shuppan, 1979.

Tichane, Robert. *Those Celadon Blues.* Painted Post, New York: New York State Glaze Institute, 1983.

_____. *Ash Glazes.* Painted Post, New York: New York State Glaze Institute, 1987. A methodical exposition of the role of ash in glazes. The discussion of soluble alkali is particularly useful.

_____. *Clay Bodies.* Painted Post, New York: New York State Glaze Institute, 1990.

Whitford, Phillip, and Gordon Wong. *Handmade Potter's Tools.* Tokyo, New York, and San Francisco: Kōdansha International, 1986. A guide for potters who want to develop their own tools and find art in them. Instructions on how to make many of the tools mentioned in this book.

Wood, Nigel. *Oriental Glazes.* London: Pitman Publishing, 1978. A compact but sophisticated account of the development of Chinese and, to some extent, Japanese glazes. Includes discussion of how these glazes may be approached with Western materials.

Index

(Romanized Japanese equivalents in parentheses)

Photo Sources

(in alphabetical order)

The author wishes to express thanks to the many institutions and individuals who assisted in securing photographs for the book. In addition to those credited below, the cooperation of Hara Reiko, Itō Tetsuo, Kobiki Harunobu, Setsu Iwao, Shiraishi Masami, Takizawa Kōji, Uchida Tokugo, and Yabe Yoshiaki is deeply appreciated.

Color

Craft Gallery, National Museum of Modern Art, Tokyo (19)
The Gotoh Museum, Tokyo (9, 11)
The Hachirō Yuasa Memorial Museum, International Christian University, Tokyo, photography by Saiki Osamu (20, 21, 22, 24, 25)
Kamei Rakuzan, Fukuoka (5)
Los Angeles County Museum of Art, General Unrestricted Acquisitions, Far Eastern Art Department Deaccession Funds, Frederick Weisman Company, and the Far Eastern Art Council (1)
MOA Museum of Art, Atami (13, 14, 16, 17, 18)
Setsu Gatōdō, Tokyo (4)
Tokyo National Museum (2, 8)

Black and White

Azuhata Shinichi (40, 42, 52, 80, 88, 95, 127)
Art Gallery, Kyoto City University of Fine Arts (18, 139)
Craft Gallery, National Museum of Modern Art, Tokyo (16)
Gotoh Museum, Tokyo (12, 132)
Idemitsu Museum of Art, Tokyo (15)
Keiō University, Tokyo (10)
Ōbata Masanori (129)
Ogawa Chōraku (138)
Seattle Art Museum, Margaret E. Fuller Purchase Fund, Asian Department Restricted Funds, Asian Art Council Purchase Funds, and Helen Seymon Memorial Fund, 85.28. Photo: Paul Macapia (9)
Setsu Gatōdō, Tokyo (11, 128, 131)
Shinji Shūmeikai, Shiga (14)
Tokyo National Museum (5, 6, 7, 8, 13)

The "weathermark" identifies this book as a production of Weatherhill, Inc., publishers of fine books on Asia and the Pacific. Editorial supervision: Jeffrey Hunter. Book and cover design: Mariana Canelo. Production Supervision: Bill Rose. Page composition: Joseph Sherman. Printing and binding: Oceanic Graphic Printing, Hong Kong. The typeface used is Scala for text and display.